Engineering Materials

This series provides topical information on innovative, structural and functional materials and composites with applications in optical, electrical, mechanical, civil, aeronautical, medical, bio- and nano-engineering. The individual volumes are complete, comprehensive monographs covering the structure, properties, manufacturing process and applications of these materials. This multidisciplinary series is devoted to professionals, students and all those interested in the latest developments in the Materials Science field, that look for a carefully selected collection of high quality review articles on their respective field of expertise.

Indexed at Compendex (2021) and Scopus (2022)

Arnold A. Lubguban · Roberto M. Malaluan ·
Gerard G. Dumancas · Arnold C. Alguno

Bio-based Foam Sorbents

For Oil Spill Cleanups

Springer

Arnold A. Lubguban
Center for Sustainable Polymers,
Department of Chemical Engineering
and Technology
Mindanao State University—Iligan Institute
of Technology
Iligan City, Philippines

Roberto M. Malaluan
Center for Sustainable Polymers,
Department of Chemical Engineering
and Technology
Mindanao State University—Iligan Institute
of Technology
Iligan City, Philippines

Gerard G. Dumancas
Department of Chemistry
North Carolina Agricultural and Technical
State University
Greensboro, NC, USA

Arnold C. Alguno
Department of Materials and Resources
Engineering and Technology
Mindanao State University—Iligan Institute
of Technology
Iligan City, Philippines

ISSN 1612-1317 ISSN 1868-1212 (electronic)
Engineering Materials
ISBN 978-981-96-1459-2 ISBN 978-981-96-1460-8 (eBook)
https://doi.org/10.1007/978-981-96-1460-8

This Springer imprint is published by the registered company Springer Nature Singapore Pte Ltd.
The registered company address is: 152 Beach Road, #21-01/04 Gateway East, Singapore 189721, Singapore

This book is dedicated to all the visionaries, researchers, scientists, and environmentalists out there passionately seeking alternative ways to effectively and efficiently remediate oil spills without causing more damage to our waters and environment. And to all those who believe that protecting and nurturing our resources will help to heal the world, may this serve as an inspiration on continuously looking for sustainable and innovative solutions to preserve the world.

Foreword

The environmental and ecological crises caused by oil spills are among the most devastating challenges facing our planet today. These incidents not only disrupt marine ecosystems but also endanger wildlife,harm coastal communities, and degrade vital natural resources. While conventional oil spill response techniques have played a significant role in containment and cleanup, they often fall short in terms of efficiency, sustainability, and long-term environmental compatibility. In this context, bio-based foam sorbents represent a transformative solution, combining eco-friendly principles with cutting-edge material innovation.

This book, Bio-Based Foam Sorbents, offers a comprehensive examination of the potential and promise of bio-based foam technologies for oil remediation. The authors have skillfully detailed the development,characterization, and performance evaluation of these materials, presenting a forward-looking perspective that aligns with global sustainability goals. Key areas covered in this work include the design and synthesis of bio-based foams, their unique structural properties, and the mechanisms underlying their high sorption capacities. The book also addresses the critical aspects of biodegradability, scalability, and cost-effectiveness,ensuring that these solutions are not only scientifically robust but also commercially viable.

What distinguishes this book is its emphasis on integrating renewable resources into advanced materials design. By utilizing bio-derived polymers, natural fibers, and other sustainable inputs, the development of these foam sorbents reflects a commitment to reducing reliance on synthetic materials and mitigating secondary environmental impacts. This innovative approach demonstrates how the valorization of waste biomass and other renewable feedstocks can contribute to a circular economy while addressing urgent environmental challenges.

For researchers, engineers, and policymakers alike, this book serves as an essential resource. It bridges the gap between academic research and real-world application, offering insights that will inspire further advancements in the field of oil spill remediation. Moreover, it underscores the importance of collaboration across disciplines to achieve practical and impactful solutions to some of the most pressing environmental problems of our time.

As the global community continues to seek sustainable answers to ecological crises, the knowledge and strategies presented in this book will undoubtedly serve as a foundation for future innovations. It is a testament to the power of scientific ingenuity and environmental responsibility working in harmony.

Dr. Randall Maples
Department Chair and Associate
Professor of Chemistry
East Central University
Ada, OK, USA

Preface

Oil spill is still one of the most devastating challenges faced by our waters and environment today. Despite the extensive effort of oil spill cleanups, enhancing and developing cleanup strategies and sustainable methods, it still remains crucial. Numerous methods are tailored for oil spill cleanups namely in-situ burning, bioremediation, mechanical recovery, and chemical dispersants; however, it is important to know their advantages and disadvantages. Aiming to find the middle ground for all these strategies, this book introduces bio-based foam sorbents for oil spill remediation. Bio-based foam sorbents in this book are foam sorbents developed with bio-based components that are able to absorb/adsorb oil and repel water. The book introduces oil spill and sorbents, along with the history of oil spill and traditional remediations (Chap. 1), discusses the chemistry behind the synthesis and materials of the bio-based sorbents, (Chap. 2) elaborates the factors affecting foam sorbent performance (Chap. 3), discussed the properties and characteristics of the bio-based foams (Chap. 4), discussed the scalability and sustainability of the foams (Chap. 5), and explored the future innovation and different applications of the bio-based foams. This book contains information about bio-based foam sorbents for oil spill cleanup that will help students, researchers, scientists, and other entities interested in innovating sustainable and eco-friendly solutions for oil spills.

Iligan City, Philippines — Arnold A. Lubguban
Iligan City, Philippines — Roberto M. Malaluan
Greensboro, USA — Gerard G. Dumancas
Iligan City, Philippines — Arnold C. Alguno

Acknowledgements We extend our heartfelt gratitude to Mindanao State University—Iligan Institute of Technology and the Center for Sustainable Polymers (CSP) staff for their unwavering support in bringing this book to fruition. Our deep appreciation also goes to our editor and author, Dr. Arnold A. Lubguban, for his invaluable expertise and guidance, which shaped this manuscript into its finest form. We are equally grateful to our publisher, Springer Nature Singapore Pte Ltd., for their support and interest in this work, providing us the opportunity to share this book with a broader audience. Special thanks go to the authors and co-authors whose contributions were instrumental to this book's success: Dr. Arnold A. Lubguban (Chap. 3), Dr. Roberto M. Malaluan (Chap. 1), Dr. Arnold C. Alguno (Chap. 4), Dr. Gerard G. Dumancas (Chap. 6), Roger Jr. G. Dingcong (Chap. 5), Mae Francy T. Lucid (Chaps. 1, 2,and 3), Jady Lee E. Amarillas (Chaps. 2 and 6), Glady May A. Mejias (Chap. 3), Shem Patrick Josh A. Laguda (Chap.4), David Jeiel G. Dabodabo (Chap. 4), and Carlo James H. Solidor (Chap. 2). Lastly, we express our gratitude to everyone who supported us throughout the process of writing this book. Your encouragement and contributions were invaluable.

Contents

Abbreviations

AR	Acerola bagasse residues
BM	Bagasse malt
BPU	Bio-polyurethane
C18-DE	Octadecyltrichlorosilane functionalized diatomaceous earth particles
CF	Carbon fiber
CGF	Calotropis gigantea fiber
CNT	Carbon nanotube
CR	Coffee husk residue
DBTDL	Dibutyltin dilaurate
DCS	Differential scanning calorimetry
DMBA	2,2-dimethylol butanoic acid
DMF	Dimethylformamide
EPS	Expanded polystyrene
FAS	Fluorinated alkyl silanes
FDTS/F-SiO2	1H, 1H, 2H, 2H-perfluorododecyltrichlorosilane
FESEM	Field emission scanning electron microscopy
F-SiC	FDTS modified SiC
FTIR	Fourier transform infrared spectroscopy
HDI	Hexamethylene diisocyanate
LLB-CF	Liquefied-larch-based carbon foam
LLB-PF	Liquefied-larch-based polymer foam
LMA	Life cycle assessment
LNPs	Lignin nanoparticles
LPUF	Lignin-based polyurethane foam
MDI	Methylene diphenyl diisocyanate
m-LNP	Modified-LNP
MTEBP	Modified tannin extract-based polyether polyol
MTPF	Modified tannin extract-based polyurethane foam
MW	Macadamia nutshell waste
NIPU	Non-isocyanate polyurethane

NMR	Nuclear magnetic resonance
NOx	Nitrogen oxides
NR	Natural rubber
NRF	Natural rubber foam
NSF	Naturally superoleophilic foam
OCC	Open cell content
OTMS	Octadecyltrimethoxysilane
PA	Phytic acid
PAAC	Modified biomass carbon fibers foam immobilized with $Ag_2O@Ag$ nanoparticles via the self-assembly, freeze-drying, carbonization, and hydrophobic modification processes
PAHs	Polycyclic aromatic hydrocarbons
PBS	Poly(butylenes succinate)
PCF	PLA-CGF foam
PDA	Polydopamine
PDMS	Polydimethylsiloxane
PEG	Polyethylene glycol
PF	PLA foam
PHA	Polyhydroxyalkanoates
PHBV	Poly(3-hydroxybutyrate-co-3-hydroxyvalerate)
PLA	Polylactic acid
p-MDI	Diphenylmethane diisocyanate
POBP	Palm-oil-based polyol
PPG	Polypropylene glycol triol
PR	Palm fiber residue
PU	Polyurethane
PUF	Polyurethane foam
PVC	Polyvinyl chloride
rGO	Reduced graphene oxide
RH	Rice husk
RHA	Rice husk ash
RPUF	Rigid polyurethane foam
SC-PU	Sugarcane bagasse-polyurethane *composite*
SEM	Scanning electron microscope
Si-C	Silicon carbide
SOx	Sulfur oxides
TBA	t-butanol
TDI	Toluene diisocyanate
TEM	Transmission electron microscope
Tg	Glass transition temperature
TGA	Thermogravimetric analysis

VER	Volume expansion ratio
WCA	Water contact angle
WCO	Waste cooking oil

Chapter 1
Introduction to Oil Spill and Sorbents

Abstract Oil spills create serious ecological and economic problems making effective cleanup methods in great demand. The objective of this chapter is to give an overall view of the problems and impacts that oil spills cause. This sets the stage for the solutions explored in this book. Several clean-up strategies applied for oil spills will be reviewed with a focus on the advantages and disadvantages of sorbents compared to other methods. This chapter further reviews the contribution of sorbents to oil spill cleanup in consideration of different types and forms, together with their intrinsic limitations. Efficiency, selectivity, environmental impact, economic feasibility, scalability, and deployment challenges related to the use of foam sorbents are critically analyzed. It also gives a short historical background on some important oil spill incidents and the use of sorbents in cleanups. It is centered on bio-based foam sorbents and emphasizes the use of sustainable methods, showing examples how they work in oil spill cleanups.

Keywords Oil spil · Sorbent · Sorption · Deep horizon · Exxon Valdez · Oil sorption capacity · Hydrophobicity · Bio-base · Oil leak · Bioremediation · Chemical dispersants · In-situ burning

1.1 How Oil Spills Affect the Environment

Oil spill is an accidental release of oil into the environment, usually in oceans and coastal areas. These spills are mainly associated with human activities such as drilling, transporting and refining oil. This problem has remained challenging up to this day, with serious consequences of oil spills, causing harm to the environment and damaging the ecosystems, which resulted in economic losses for the communities that depend on fishing and tourism [1, 2]. The two main causes of oil spill include human action and natural occurrences. Oil spills caused by human activities often happen during the transportation of oil where oil tankers collide or pipelines break. Oil can also leak accidentally from ships during routing operations or maintenance, and old pipelines and storage tanks may also develop leaks, particularly in regions

A. A. Lubguban et al., *Bio-based Foam Sorbents*, Engineering Materials,
https://doi.org/10.1007/978-981-96-1460-8_1

with extensive oil extraction activities [3]. On the other hand, natural oil spills can occur on the ocean floor from oil seeps rising through the surface, although these types are typically smaller in size compared to human-related spills [4].

The higher global demand for oil, coupled with increased oil exploration, has heightened the prospect of a spill, in view of the fact that more oil is traveling around ecologically sensitive areas than ever before [5]. Places with extreme weather conditions or places that majorly produce oil like the Arctic and the Niger Delta face special risks regarding oil pollution through operational and accidental spills. Bad weather and sea currents spread the oil in no time, making any cleaning-up work difficult to conduct, and prolonging damage to the environment [6].

The immediate environmental effects of an oil spill are usually catastrophic. Oil can coat marine life which causes the death of fishes, birds, and marine mammals by smothering or poisoning or loss of insulation [7]. Even exposure to oil has been reported to disrupt the circulatory systems in fish and other aquatic animals, leading to population decline [8]. High concentrations of oil can also lead to the collapse of phytoplankton communities, critical to marine food webs, and thus to the total destruction of an entire ecosystem [7]. The potential long-term ecological impacts of oil spills may take a number of years to even decades. Comparisons with major spills, such as Exxon Valdez and Deepwater Horizon, have suggested that recoveries are really long, around 2–10 years, though some species and habitats will continue to be changed far beyond that initial cleanup phase. Long-lasting pollutants like PAHs (polycyclic aromatic hydrocarbons) may continue to burrow into sediment-dwelling organisms and disturb community structures through reduced biodiversity.

Economically, the impacts of oil spills are too high and most communities heavily depend on fish, tourism, and other marine resources. These recorded significant losses resulting from spilled oil from contaminated fishing sites and low tourist activities [9, 10]. The economic costs of oil spills are usually huge, in the form of cleaning costs, income losses by stakeholders in industry, and other related long-term health defects among the native people [11]. In addition, oil spills can warrant social strife and conflicts over resources, especially in those communities that feel that their livelihood is threatened due to the negligence of a corporation or laxity in regulation [10]. In essence, the psychological consequences that flow from communities through the loss of livelihood and depreciation of the environment are likely to be enormous and affect their entire well-being [9].

On a broader level, oil spills highlight the significant risks associated with large-scale dependence on fossil fuels. Spills show long-term effects, specifically on climate change, and the collapse of natural environments. The consumption and burning of fossil fuels contribute a great deal in the emission of greenhouse gases, while oil spills as such reveal the risks to the environment from these sources of energy. That is, there is more concern with oil spills today, which at times is a direct consequence of the increased number of vessels at sea and the number of offshore drills, in light of climate change. In such a way, while temperatures are rising and patterns of weather shift, ecosystems will be further weakened to effectively respond to oil spills, thereby suffering more serious and long-term impact [8].

Under the circumstances, there is an increasing demand for effective and practical techniques in oil spill response. The book presents the possibilities of bio-based foam sorbents to be employed as feasible options to the traditional sorbents for oil spills. The aim is to show that the bio-based foam sorbents can prove to be competitive with the existing commercial products in relation to their use and applicability. Another key aspect of this book is the evaluation of all the different factors that affect the actual performances of these foam sorbents such as material constitutions, physical and chemical characteristics, and environmental conditions. Another key aspect dealt with in the book is the economic viability of the bio-based foam sorbents with a view toward their commercialization. It views sustainability and cost-effectiveness as key ingredients for success upon large-scale adoption. With this, the book sets out to prove that indeed, the bio-based foam sorbents could offer a sustainable solution to oil spill remediation in addressing global demands for environmentally friendly and efficient methods of mitigating oil spill impacts.

1.2 Oil Spill Remediations

Over time, various methods have been developed for oil spill remediation that can reduce the level of environmental devastation because of accidental releases of oil. The techniques used in these different methods vary with regard to approach, efficiency, and environmental impact thus careful consideration in terms of their relative advantages and disadvantages becomes essential in choosing an appropriate method for a given situation.

The most common ones are mechanical containment and recovery, dispersals, and bioremediation. Mechanical containment and recovery includes the use of booms, skimmers, and barriers. Booms are floating barriers that prevent the dispersion of the spill to spread further away from the area where it has happened while skimmed are used to take oil out of the water surface. The main positive side of mechanical methods consists in the fact that they can contain and recover large amounts of oil, especially in placid waters [12]. Besides, all these methods are less harmful to the environment than chemical ones because they do not introduce anything foreign into the ecosystem [13]. However, they also possess some drawbacks. First, mechanical containments are highly dependent upon weather conditions, with their efficacy related to wind and wave actions, including currents that may easily disrupt the containment efforts [14]. On the other hand, most of these mechanical operations of recovery are believed to be expensive and time-consuming, moreover, many of them require highly substantial logistics support and specialized equipment too [15].

Another of the most commonly used methods is the use of chemical dispersants. Dispersants are the kind of chemicals that succeed in breaking the oil down into smaller droplets that finally join in the water easily. This is a process that in turn helps to reduce the surface slick and thus the microorganisms get on quickly with the process of breaking down material. The primary benefits of dispersants are their quick performance in the case of large spills and their capacity to stop most of the oil from immediately coating sensitive shorelines and marine life [16]. On the other hand, the use of chemical dispersants has been the subject of intense discussions as a result of the environmental and health risks they potentially can pose. Dispersants can lead to the rendering of oil poisonous to the marine organisms, and in this way, they can become more of a toxic substance. Their efficiency is very much reduced in some specific conditions such as a condition with low temperature or a condition with a lot of sediments in it [17].

Besides the fact that oil is mainly decomposed to less harmful chemical substances through mechanical and chemical methods, bioremediation uses microorganisms to do the job naturally. The eco-friendly technology, bioremediation, enables a natural decomposition of the oil without burning any chemicals into the atmosphere [18]. Sustainability and treatment of oil-contaminated areas have the advantage of bioremediation over a longer time frame [19]. However, bioremediation is a slow and static technology that may not be suitable for the immediate response to large-scale oil spills. Its success is similarly affected by environmental factors such as temperature, nutrient availability, and oxygen levels, where they cannot support the microorganism's oil breakdown. Such factors might as well restrict the operation of the oil-degrading microorganisms resulting in lower efficiency [20].

Of the many methods, sorbents are critical in oil spill cleanup. Sorbents are materials capable of absorbing or adsorbing oil, thus effectively taking it out from water. They are very instrumental for small-scale spills or conditions where more invasive methods may further be detrimental to the environment [21]. Sorbents may be of natural, synthetic, or a combination of both. Their field performance varies by their composition and design. The simplicity of sorbents is one of their main advantages, as they can be quickly deployed in emergency situations to prevent further spread of the oil [22]. However, their capacity is limited and they are usually changed frequently, hence many times not economically viable for big spills.

One of the innovations in this area is the foam sorbent, in particular those which are obtained from natural materials such as coconut oil-based polyol. On the other hand, foam sorbents have some distinct features that make them much better than traditional sorbents. They have a great capacity to take in oil more than their weight in some cases. This high absorption capacity makes foam sorbents particularly valuable for both large-scale spills and smaller, more contained incidents. In addition, foam materials can be inherently hydrophobic or made to be very hydrophobic, meaning they selectively absorb oil while repelling water, which is crucial in the context of oil spill cases in marine areas. In addition to the aforementioned benefits, foam sorbents can be reprocessed, cleaned, and reused many times without any significant drop in the sorption capacity among other residual waste. They are also known to have a minimal environmental impact when compared to one-time use sorbents.

Each of these methods has its place in an integrated oil spill response strategy compared to other remediation techniques. Mechanical, chemical, and biological methods are often favored for larger spills as they can handle substantial volumes of oil. By comparison, the use of sorbents, especially foam sorbents, is best suited for point applications-confined spaces or environments that have high ecological value. However, the selection of any method is controlled by a number of factors that include size and location of the spill, environmental sensitivity, and urgency of the response. The other unique advantage of the foam sorbents pertains to environmental impact, since it can be designed to be biodegradable and friendly to the environment, unlike some chemical dispersants, which may introduce other pollutants into the environment.

1.3 Sorbents

Sorbents are those types of materials that can absorb or adsorb liquids or gases from any other substance by absorption or adsorption. In oil spill cleanup, the sorbent is specifically designed to recover oil from water by absorption, where the oil permeates the entire structure of the material, or adsorption, where it clings to the surface of the sorbent [23]. The effectiveness of the sorbents refers to the degree at which a material can absorb oil while resisting water uptake, which is of huge importance when one is trying to reduce environmental damage related to an oil spill [14].

Because of their capacity for swift and efficient removal of oil from the water surface, preventing further spread or environmental damage, sorbents remain one of the mainstays of oil spill response. In general, sorbents can only be applied in sensitive areas where other methods of mechanical recovery either do not apply or could be destructive, such as marshlands, shorelines, or biodiversity hotspots [12]. For this reason, proper selection of sorbents will be able to minimize possible environmental and economic impacts of an oil spill which is an integral part of any oil spill response strategy [16].

In general, there are three types of sorbents shown in Table 1.1: natural organic, natural inorganic, and synthetic. The first ones are straw, peat moss, and coconut husk. Natural organic sorbents are biodegradable and environmentally friendly but might have relatively lower oil sorption capacity and may be prone to water absorption. Natural inorganic sorbents are generally of high sorption capacity and include clays and perlites, however, they may also be heavy and difficult to deploy in any large-scale spill cleanup [37]. On the contrary, artificial sorbents made from polymeric materials such as polypropylene are characterized by high oil sorption capacity and relatively low water intake and, therefore, can serve as very efficient oil spill cleaners. They also create an environmental hazard problem, as they are non-biodegradable and may be a considerable source of microplastic pollution [17].

The forms of sorbents include pads, booms, and loose particulate materials in the form of foams, each for a particular use based on various conditions of spills. Pads are typically applied on small spills or surface oil, while booms are deployed on open water to contain and absorb the oil. Loose particulate materials are applied on those areas where the oil has already percolated into the ground or on those areas where precision applications are needed. Generally, the sorbents are those highly absorbent materials made from either synthetic or natural polymers. By nature, they have a structure that is porous, enabling them to hold or retain large volumes of oil in relation to their weight. This then makes them very ideal for large-scale spills or any other

Table 1.1 Types of sorbents

Type of sorbent	Example material	Application	Characteristics
Natural organic	Straw	Small spills, marshlands, coastal areas	Biodegradable, locally available, limited capacity [24]
	Peat moss	Small spills, sensitive wetlands	High oil absorption, hydrophobic, environmentally friendly [25]
	Coconut husk	Shoreline and sensitive habitats	Biodegradable, cost-effective, moderate capacity [26]
Natural Inorganic	Clay	Industrial areas, soil remediation	High sorption, non-biodegradable, dense and hard to handle [27]
	Perlite	Soil and surface spills	Light, high capacity, chemically inert, not suited for marine spills [24]
	Vermiculite	Soil, construction spill management	Lightweight, stable, chemically inert, cost-effective but less effective in open water [28]
Synthetic polymer	Polypropylene booms and pads	Large-scale marine spills, surface slicks	High sorption, reusable, hydrophobic, creates microplastic pollution [29]
	Polyurethane foam	Marine and inland waters, rapid deployment	High capacity, reusable, oleophilic and hydrophobic, durability issues in harsh environments [30]
	Acrylic fibers	Small-scale, targeted spills in marine environments	Absorbent, hydrophobic, non-biodegradable, suitable for small spills [31]
Bio-based foam	Coconut oil polyol-based foam	Coastal, wetland areas, small to medium spills	Biodegradable, renewable source, high sorption, limited durability [32]
	Cellulose-based sorbents	Coastal areas, marshes, small spills	Biodegradable, high absorption, eco-friendly, limited reuse [33]

(continued)

Table 1.1 (continued)

Type of sorbent	Example material	Application	Characteristics
Advanced materials	Graphene-based sorbents	High-precision applications, experimental	Extremely high surface area, oleophilic, lightweight, expensive [34]
	Chitosan	Targeted spill response in ecologically sensitive areas	Biodegradable, renewable, antimicrobial properties, costly [35]
	Silica aerogels	Laboratory spills, experimental marine applications	Super-absorbent, lightweight, hydrophobic, still in experimental stages for large-scale use [36]

situation where high capacity for absorption is of essence. However, even with these excellent performance results of these sorbents, there are several disadvantages that still exist such as saturation capacity, possible secondary contamination, and difficult recovery and disposal after use. Moreover, environmental conditions, oil type, and specific demands related to the spill require careful selection for ideal performance [38].

In these forms of sorbents, the foam-based sorbents are superior, having very high porosity and large surface area with lightweight, hence better oil absorbency and handling. Specifically, polyurethane or other synthetic polymer foams can absorb many times their weight in oil and hence become effective in addressing large spills accordingly [19]. Besides, the sorbents can also be designed to have properties of being oleophilic and hydrophobic, hence enhancing its selectivity toward oil in water [39]. The improved properties generally limit the risk of water intake and improve the efficiency of the clean-up exercise.

Recent design of the bio-based foam sorbent from renewable resources such as polyol-based coconut oil has brought the foam into the spotlight as more ecologically friendly options for oil spill clean-ups. These foams exhibit the advantages of the conventional synthetic foams along with enhanced environmental compatibility, high oil sorption capacity, and recyclability [40].

1.4 Challenges in Choosing Sorbents for Oil Spills

One of the main challenges for the application of foam sorbents in oil spill cleanup is the optimization of their efficiency and selectivity. Efficiency in this respect is defined by the oil sorption capacity which is the amount of oil which can be absorbed by a certain amount of sorbent relative to its weight. In order to successfully apply the foam sorbents to handle large-scale oil spills efficiently, it is important that the former exhibit high oil absorption efficiencies. It is closely associated with efficiency factors such as porous structure, surface area, and chemical properties that would be favorable for oil sorption [41]. On the other hand, efficiency may be hard to sustain, especially under varying environmental conditions that may alter some properties of oil and dynamics of spills.

Aside from efficiency, another aspect that is equally important is selectivity. In practice, sorbents need to differentiate between oil and water by absorbing oil and repelling water in order to function effectively in realistic scenarios, especially in marine environments where water is always present. Hydrophobicity shown by its water contact angle plays an important role in selectivity of the foam sorbents. A high water-contact angle indicates better repellency of water and, consequently, higher oil selectivity [19]. At any rate, such high selectivity without sacrificing other performance characteristics has yet to be a formidable challenge. This problem is especially the case with conventional petroleum-based sorbents, which might need further chemical treatment in order to improve their selectivity.

Another major concern in assessing sorbent use in oil spill cleanup would be environmental impact. Conventionally, the foam sorbents often require complicated chemical treatments to improve their performance where some of these include toxic solvents. These chemical treatments not only pose an environmental hazard due to the use of harmful chemicals but also because of long-term environmental contamination during the disposal of such sorbents. Another risk posed is because of their non-biodegradable nature in which sorbents may accumulate in ecosystems and further threaten wildlife and habitats [42]. All the foregoing environmental drawbacks associated with petroleum-based foams have been driving the search for more environmentally friendly alternative products.

This bio-based foam sorbent from renewable feedstock, including vegetable oils and biomass, offers the same potential for environmental mitigation. These bio-based materials reduce dependence on the non-renewable petroleum products and offer a much-reduced carbon footprint [43]. Also, their fabrication usually involves fewer toxic chemicals to meet the environmental safety goals. Finally, bio-based foams can be recycled many times with little performance degradation which reduces costs and waste, making them more sustainable and economically feasible solutions for oil spill response [44].

Financial viability also is an important determining factor in the choice of sorbent selection for the cleanup of oil spills. For example, the manufacturing and disposal of petroleum-based sorbents are expensive, since compliance with environmental regulations involved in such operations and its management of waste will be very costly [42]. Finding cost-effective sorbents that do not compromise on performance or environmental safety continues to drive research and development of this subject.

The economic impacts of deploying sorbents, beyond initial costs, are a function of their net sorption performance. The more highly absorbing sorbents will decrease the quantity of material necessary for effective cleanups, decreasing both the material and waste disposal costs. Many of the foam sorbents can also be reused many times, which furthers their long-term economies [21]. In spite of these advantages, one again it is a challenge to develop sorbents that are economically viable and environment-friendly.

Another factor that will determine practical applicability of the application of the foam sorbents into a wider practice will be scalability of their production. For large-scale viability, sorbents should be producible in considerable amounts over short periods of time due to the sudden nature of spill events. Preferably, the manufacturing process would not be overly complicated or require special facilities beyond those already used in industry to facilitate rapid production. Whereas the one-shot foaming process, among other techniques, is in mainstream use, it remains cognizant that scaling up production without losing product quality or performance cannot be done consistently [45].

Finally, deployment strategies are of great import for the real-world applications of these foam sorbents. These sorbents should be lightweight and easy to handle for easy storage and transportation for application across a wide range of spill sites, most of which are rather challenging in their nature, especially in marine environments. Their structural integrity should resist mechanical stresses during deployment and recovery operations. Research in both petroleum-based and bio-based foams is continuing to make these foams more durable, reusable, and easily deployable [46]. Parallel to that, the development of application and recovery machinery in the case of deploying the use of the foam sorbents should be carried out in order to be effectively employed for large-scale and repetitive operations [15].

1.5 Historical Background

Oil spills have been an environmental concern over the last hundred or so years. Some of the earliest recorded oil spills were in the early twentieth century. One of the first large oil spills was the Lakeview Gusher in California in 1910; an estimated 9 million barrels of crude oil spilled into the environment [47]. Primarily an onshore incident, it showed the devastating result of uncontrolled oil release. Within the mid-twentieth century, oil production and transportation have grown in scale, even more so than accidents or incidents of oil spills. A number of notable events since the Torrey Canyon spill in 1967-a supertanker ran aground off the English Coast, spilling approximately 100,000 tons of crude oil-raised awareness globally regarding the environmental impact associated with oil spills and the need for effective response strategies [19].

Perhaps one of the most infamous incidents in recent history occurred with the Exxon Valdez in 1989. This incident was due to an oil tanker striking a reef off the coast of Alaska released gallons of crude oil which is approximately 11 million gallons into Prince William Sound. This environmental catastrophe destroyed more than 1300 miles of shoreline and brought concrete ramifications of oil spills. As a result, this event led to further policy and technological shifts in how oil spills would be prevented and addressed [48]. Another incident was in 2010 where a very serious oil spill was considered one of the largest offshore oil spills experienced at this point in time. The accident recorded more than 200 gallons spilled crude oil, causing huge destruction to marine and coastal ecosystems into the Gulf of Mexico. It once more captured attention that the problem of management of oil spillages was far from being solved, at least in deep-water drilling [49].

Various methods for cleaning up oil spills have, therefore, been developed in response to the incidents which range from mechanical recovery using skimmers and booms to chemical dispersants that break down oil into minute droplets. However, all these methods had their own limitations regarding efficacy as well as environmental concern, which gave rise to the need to look for alternatives. From the 1960s to the 1970s, much interest was aroused in sorbents, materials which absorb or adsorb liquids, as a viable method for oil spill response [42].

Naturally abundant sorbents, such as straw, sawdust, and peat moss, were initially used because of their easy availability at low cost. These showed some oil sorption capability, but their low oil sorption capacity, quick saturation, and other handling and disposal problems prevented their extensive usage in this area. Application of natural sorbents for oil absorption was done onshore, but their efficiency was small due to rapid saturation and poor durability in harsh marine conditions [12]. These disadvantages pointed out a requirement for more effective sorbents in that field to handle large-scale oil spill incidents.

These limitations gave rise to synthetic sorbents, mostly derived from petroleum resources such as polypropylene and polyurethane foams. Synthetic sorbents have higher oil sorption capacity, hydrophobicity, and durability compared to natural sorbents and are preferred during large-scale oil spill response operations. The synthetic sorbents gained prominence in the cleaning process of the Exxon Valdez spill, where thousands of polypropylene pads and booms were used to absorb and contain the spread of oil. The use of such materials gave a much-required fillip in effectiveness with respect to oil removal from water surfaces in a real spill situation [50].

In the 1990s, increased sophistication in sorbent technology saw the introduction of specially designed sorbent products made from nonwoven polypropylene fibers and mats, and booms that applied frequently and widely during oil spill responses. These became lightweight, easy to handle, and much more endowed with higher absorbing capabilities while at the same time allowing water to repel. Their hydrophobic and oleophilic characteristics enabled them to float on the water surface and thus had great applications in skimming operations and shoreline cleanup [51]. An example of the applicability of synthetic sorbents was during the Prestige oil spill off the coast of Spain in 2002, where it played a very important role in containing the spread into sensitive coastal environments by containing and absorbing it before it reached coastlines [21].

In application, sorbents have played their vital role in many cases of oil spills, with each having its results or lessons to be learnt about their applicability and effectiveness. During the Exxon Valdez spill, synthetic sorbents became the handy sorbent material used for cleaning both onshore and offshore spills. In this incident, polypropylene sorbent pads, rolls, and booms were used to absorb oil from the water surface and also to protect sensitive coastal habitats. While these sorbents have been able to take out sizable quantities of oil, their success underlined the practicability of these materials in large-scale cleanup operations [52]. However, its environmental viability is in question since most were usually incinerated or sent to landfills after use [53].

Deepwater Horizon presented a unique set of challenges because the discharge of oil was from a deep-sea source. Besides the conventional methods of containment and recovery, several types of sorbents have been tried and deployed in capturing huge amounts of released oil. The oil was surrounded and contained by absorbent booms of synthetic fibers, while sorbent pads and pom-poms rencont some oil was removed from the water surface and off shorelines [16]. In response to the scale of the disaster, innovative sorbent materials were trialed, including those enhanced with chemical treatments designed to increase hydrophobicity. Among these sorbents, some had very high oil uptake capacity, while others presented problems regarding deployment logistics or recovery efficiency [19].

In 2009, during the cleanup operations after the Montara oil spill in the Timor Sea, sorbents played a very important role. In this context, polyurethane foam sorbents were used in large quantities to absorb the oil from the surface of the water, as these had been designed with high oil-absorption capacity and high durability in order to reduce the impact of the spill on marine life and coastal environments [15]. These sorbents were selected based on the harsh conditions they would encounter in the open sea, including UV radiation and saltwater that could degrade less robust materials [19].

Recognizing the environmental disadvantages of classical synthetic sorbents, in recent years more and more attention has been paid to the development of bio-based sorbents manufactured from natural renewable materials. Such materials provide a promising alternative to conventional petroleum-based sorbents because of their low dependence on fossil fuels and biodegradable characteristics. For example, sorbents prepared from cellulose, cotton, wool, and other natural fibers were studied with respect to oil sorption properties along with environmental advantages [54].

Bio-based foam sorbents represent one of the key developments of recent years. Such foams, prepared from renewable feedstocks like vegetable oils, by-products of agricultural activities, and biomass, usually demonstrate very high oil sorption capacities and a degree of hydrophobization as good as their synthetic counterparts [55]. Thus, the innovation of polyurethane foam based on the incorporation of coconut oil-based polyols has appeared very promising, due to the fact that such materials inherently contain hydrophobic properties and excellent recyclability that may lead to less harm to the environment [42]. These are developments complemented by ongoing research in the chemical modification of natural polymers to improve their oil sorption capacities and to make them more durable [19].

Recent oil spill incidents also inspired further advances in sorbent technology. Following the Deepwater Horizon spill, concerted research was conducted toward more effective and sustainable sorbents. Studies were done with chitosan which is a biopolymer derived from crustacean shell waste-because of its good oil sorption capacity and biodegradability [42]. Besides, sorbents based on graphene with high surface area and good adsorption capacity were also tested for possible application in oil spill cleaning processes [43]. These materials are still under research and development but surely represent a fast-growing trend in tackling environmental challenges using advanced approaches of material science [44].

Because of that, very fast implementation of sorbents for the clean-up operation of oil spillage demands careful consideration of various factors that include effectiveness and their feasibility in terms of the environment and economics. Traditionally used synthetic sorbents work under operational use, but their post-use disposal and environmental footprints are threatening. In this regard, bio-based sorbents seem to be a good feasible alternative, but scalability challenges must be overcome along with further reduction in operational cost and performance consistency across a wide range of environmental conditions [21]. There are also some logistical issues related to storage, transport, and rapid deployment of sorbents during the field operation. In any strategy of deploying sorbents, optimization will be required to achieve maximum area of operation with high efficiency in oil absorption. Development of specific equipment for deploying and retrieving sorbents, like sorbent application vessels and vacuum recovery systems, becomes an important element of effective spill response.

1.6 Sustainable Solutions for Oil Spills

Growing environmental awareness and the strictness of regulations in pursuit of protection for marine ecosystems have raised a pressing demand for sustainable solutions in oil spill cleanup. Traditional methodologies for addressing oil spills involve chemical dispersants and petroleum-based sorbents, normally accompanied by a great deal of environmental compromise in toxicity and non-biodegradability. Such challenges have necessitated interest in developing bio-based foam sorbents as eco-friendly alternatives. The utilization of bio-based sorbents, produced from renewable feedstocks like vegetable oils, by-products in agriculture, and natural fibers, therefore represents a very viable sustainable pathway toward environmental conservation [47].

The concept of the bio-based foam sorbents would be to unite the high oil absorption capacity with environmental safety. Commonly, they possess inherent hydrophobic-oleophilic characteristics, thus ensuring selectivity in oil absorption with the repellence of water. This indeed is very useful in marine environments where repellency of water is crucial for successful oil spill response. Moreover, it is possible to design bio-based foams highly porous with high surface areas for enhanced oil absorption efficiency [56]. Furthermore, the addition of polyols based on vegetable oils in polyurethane foams increases their hydrophobicity and oil sorption capability, hence, this class of polyurethane foams emerges as an efficient way to clean up oil spills in an eco-friendly manner.

The advantages related to bio-based foam sorbents extend beyond their immediate application. Bio-based sorbents, unlike conventional synthetic sorbents that typically go into landfills or incineration post-use, are biodegradable, hence decreasing the long-term impact associated with waste disposal [47]. Biodegradation decreases the risk of secondary pollution processes and ensures that clean-up activities do not contribute to further environmental degradation. Moreover, the carbon footprint with regards to oil spill response would be further reduced by the use of the bio-based materials, mainly because of the reduced reliance on fossil fuels as well as other poisonous gas emissions. This makes the use of bio-based foam sorbents not only practical for oil spills but also very vital in efforts toward sustainable environmental practices.

The research into sustainable oil spill cleanup solutions is not new and has been pursued and applied over the years in various ways. One of the first examples of a sustainable solution was when, in the mid-twentieth century, natural sorbents like straw, sawdust, and peat moss were used for the absorption of oil from the surface of the water in the case of incidents of small-scale spills. This has proved effective, as these materials are easy to access and biodegradable. These natural sorbents had some effectiveness, but their low absorption capacity and handling difficulties made them applicable for minor spills only [48]. Nevertheless, they have formed the foundation in developing more effective sustainable sorbent materials.

Another major breakthrough in sustainable oil spill response included the utilization of sorbents made from recycled materials. Such includes the rampant interest in the utilization of recycled waste from cotton as an oil sorbent in the late twentieth century. Cotton is highly absorbent, naturally hydrophobic, and thus would offer a biodegradable and cheap sorbent material compared to sorbents resulting from synthetic processes. Recycled cotton sorbents have been tested under different oil spill scenarios. They showed similar oil absorption capability compared to their synthetic counterpart, with added value from an environmental point of view [48]. The technology provided not only an active cleanup method but also solved one of the problems involving textile waste within the circular economy.

The Deepwater Horizon oil spill triggered a new era in the search for sustainable sorbent technologies. Agricultural by-products, such as rice husks, coconut coir, and kenaf fibers, were considered after the spill as base materials for sorbents by researchers and environmental organizations. Biobased sorbents are prepared from renewable agricultural feedstock and offer a low-cost, environmentally friendly alternative for oil spill response. Different studies conducted following the spill demonstrated that such materials had high oil sorption capacities and stayed afloat while doing so, making them practical for the marine environment, too [49]. Agricultural by-products, in addition to providing an efficient sorbent material, also gave new perspectives concerning the exploitation of waste and thus contributed to sustainable agriculture.

Recent developments have considerably opened up more methods for sustainable oil spill cleanup techniques. The most recent development in the series of sustainable sorbents is represented by the bio-based polyurethane foams manufactured using renewable polyols made from vegetable oils and some other natural feedstocks [56]. Such foams combine the desirable properties of traditional synthetic sorbents, namely high oil absorption capacity and durability, with environmental benefits related to biodegradability and reduced toxicity. Developments in the chemical modification of natural polymers, including chitosan and cellulose, have also prepared sorbents with advanced oil sorption capacities and water repellency. Such bio-based materials could grant scalable, cost-effective, and ecologically responsible solutions with the implementation of advanced fabrication techniques for oil spill clean-up.

This perspective considers the development of all other eco-friendly materials-including bio-based foam sorbents-a milestone in oil spill response. Besides, continuing growth in demand for a solution calls for research and innovation into bio-based sorbents that could assure both effectiveness in cleaning oil spills and sustainability relevant to goals of development. Application of these materials in oil spill response contributes not only to solving the immediate problems of oil pollution but also to a commitment of wide environmental stewardship and sustainability.

References

1. Feng, Q., An, C.J., Cao, Y., Chen, Z., Owens, E., Taylor, E., Wang, Z., Saad, E.A.: An analysis of selected oil spill case studies on the shorelines of Canada. J. Environ. Inform. Lett. (2021). https://doi.org/10.3808/jeil.202100052
2. Jernelöv, A.: The threats from oil spills: now, then, and in the future. Ambio **39**(5–6), 353–366 (2010). https://doi.org/10.1007/s13280-010-0085-5
3. Barron, M.G.: Ecological impacts of the Deepwater Horizon oil spill: implications for immunotoxicity. Toxicol. Pathol. **40**(2), 315–320 (2011). https://doi.org/10.1177/0192623311428474
4. Burgherr, P.: In-depth analysis of accidental oil spills from tankers in the context of global spill trends from all sources. J. Hazard. Mater. **140**(1–2), 245–256 (2007). https://doi.org/10.1016/j.jhazmat.2006.07.030
5. Geng, H., Jiang, X., Wang, D., Li, J.: Prediction and assessment of marine pollution risk from ships based on statistical analysis probability model. Am. J. Traffic Transp. Eng. **6**(2), 34 (2021). https://doi.org/10.11648/j.ajtte.20210602.11
6. Alves, T.M., Kokinou, E., Zodiatis, G.: A three-step model to assess shoreline and offshore susceptibility to oil spills: the South Aegean (Crete) as an analogue for confined marine basins. Mar. Pollut. Bull. **86**(1–2), 443–457 (2014). https://doi.org/10.1016/j.marpolbul.2014.06.034
7. Zhang, Z., Sun, H., Guo, Y.: The impact of marine oil spills on the ecosystem. Int. J. Eng. Sci. Technol. **2**(1) (2024). https://doi.org/10.58531/ijest/2/1/1
8. Barron, M.G., Vivian, D.N., Heintz, R.A., Yim, U.H.: Long-term ecological impacts from oil spills: comparison of Exxon Valdez, Hebei Spirit, and Deepwater Horizon. Environ. Sci. Technol. **54**(11), 6456–6467 (2020). https://doi.org/10.1021/acs.est.9b05020
9. Sj, O., Cc, I., Pa, O., Lc, O., Ai, H.: Crude oil and refined products spill incidence impact evaluation in Eleme; Rivers State, Nigeria. J. Energy Res. Rev. **16**(4), 38–43 (2024). https://doi.org/10.9734/jenrr/2024/v16i4346
10. Wekpe, V.O., Whitworth, M., Baily, B.: Where will the next oil spill incident in the Niger Delta region of Nigeria occur? Environ. Res. Commun. (2024). https://doi.org/10.1088/2515-7620/ad29b5
11. Negreiros, A.C.S.V., Lins, I.D., Maior, C.B.S., Moura, M.D.C.: Automated detection of oil spills in images: combining a novel feature extraction technique based on the q-exponential distribution with machine learning models. Res. Square (2022). https://doi.org/10.21203/rs.3.rs-2263261/v1
12. Kim, H., Zhang, G., Chung, T.C.M., Nam, C.: A role for newly developed sorbents in remediating large-scale oil spills: reviewing recent advances and beyond. Adv. Sustain. Syst. **6**(1) (2021). https://doi.org/10.1002/adsu.202100211

13. Dave, N.: Remediation technologies for Marine Oil Spills: a critical review and comparative analysis. Am. J. Environ. Sci. **7**(5), 424–440 (2011). https://doi.org/10.3844/ajessp.2011.424.440
14. Zhu, H., Qiu, S., Jiang, W., Wu, D., Zhang, C.: Evaluation of electrospun polyvinyl chloride/polystyrene fibers as sorbent materials for oil spill cleanup. Environ. Sci. Technol. **45**(10), 4527–4531 (2011). https://doi.org/10.1021/es2002343
15. Herren, B., Saha, M.C., Altan, M.C., Liu, Y.: Funnel-shaped floating vessel oil skimmer with joule heating sorption functionality. Polymers **14**(11), 2269 (2022). https://doi.org/10.3390/polym14112269
16. Naser, M.I.: Expended polystyrene (EPS) waste as sorbent for crude oil spill cleanup: from laboratory experiments to field application. J. Petrol. Res. Stud. **7**(1), 258–272 (2021). https://doi.org/10.52716/jprs.v7i1.180
17. Anwana, E.E., Ewemoje, O.E.: Oil sorption performance of sorbent materials examined under static and dynamic conditions. Eur. J. Eng. Technol. Res. **6**(5), 107–110 (2021). https://doi.org/10.24018/ejeng.2021.6.5.2490
18. Kathir, K., Carroll, M.: From trash to treasure: a sustainable approach to oil spill clean-up. J. Emerg. Investig. (2023). https://doi.org/10.59720/21-018
19. Li, H., et al.: Polypropylene fibers fabricated via a needleless melt-electrospinning device for marine oil-spill cleanup. J. Appl. Polym. Sci. **131**(7) (2013). https://doi.org/10.1002/app.40080
20. Aracil-Gisbert, S.,, Torregrosa-Crespo, J., Martínez-Espinosa, R.M.: Recent trend on bioremediation of polluted salty soils and waters using haloarchaea. InTech eBooks (2018). https://doi.org/10.5772/intechopen.70802
21. Ge, J., et al.: Advanced sorbents for Oil-Spill Cleanup: recent advances and future perspectives. Adv. Mater. **28**(47), 10459–10490 (2016). https://doi.org/10.1002/adma.201601812
22. Patowary, M., Pathak, K.: Facile synthesis of super hydrophobic material for selective removal of spilled-oil from water surfaces. Green Chem. Technol. Lett. **2**(2), 74–77 (2016). https://doi.org/10.18510/gctl.2016.224
23. Radetić, M.M., Jocić, D.M., Jovančić, P.M., Petrović, Z.L., Thomas, H.F.: Recycled wool-based nonwoven material as an oil sorbent. Environ. Sci. Technol. **37**(5), 1008–1012 (2003). https://doi.org/10.1021/es0201303
24. Teas, C., et al.: Investigation of the effectiveness of absorbent materials in oil spills clean-up. Desalination **140**(3), 259–264 (2001)
25. Pandey, S., Alam, A.: Peat moss: a hyper-sorbent for oil spill cleanup—a review. Plant Sci. Today **6**(4), 416–419 (2019). https://doi.org/10.14719/pst.2019.6.4.586
26. Ifelebuegu, A., Momoh, Z.: An evaluation of the adsorptive properties of coconut husk for oil spill cleanup. In: Proceedings of World Congress on Engineering and Computer Science (2015). https://doi.org/10.15224/978-1-63248-040-8-38
27. Frost, R., Carmody, O., Xi, Y., Kokot, S.: Adsorption of hydrocarbons on organo-clays—implications for oil spill remediation. J. Colloid Interface Sci. **305**(1), 17–24 (2007)
28. Sharma, K., et al.: Advancements in natural remediation management techniques for oil spills: challenges, innovations, and future directions. Environ. Pollut. Manag. **1**, 128–146 (2024). https://doi.org/10.1016/j.epm.2024.08.003
29. Bhardwaj, N., Bhaskarwar, A.N.: A review on sorbent devices for oil-spill control. Environ. Pollut. **243**, 1758–1771 (2018). https://doi.org/10.1016/j.envpol.2018.09.141
30. Zimmermann, M.V.G., Zattera, A.J., Fenner, B.R., Santana, R.M.C.: Sorbent system based on organosilane-coated polyurethane foam for oil spill clean up. Polym. Bull. **78**(3), 1423–1440 (2020). https://doi.org/10.1007/s00289-020-03169-5
31. Zhao, J., Xiao, C., Feng, Y., Xu, N.: A review: polymethacrylate fibers as oil absorbents. Polym. Rev. **53**(4), 527–545 (2013). https://doi.org/10.1080/15583724.2013.828749
32. Tomon, T.R.B., Omisol, C.J.M., Aguinid, B.J.M., et al.: A novel naturally superoleophilic coconut oil-based foam with inherent hydrophobic properties for oil and grease sorption. Sci. Rep. **14**, 14223 (2024). https://doi.org/10.1038/s41598-024-64178-2
33. Olivito, F., Algieri, V., Jiritano, A., Tallarida, M.A., Costanzo, P., Maiuolo, L., De Nino, A.: Bio-based polyurethane foams for the removal of petroleum-derived pollutants: sorption in

batch and in continuous-flow. Polymers **15**(7), 1785 (2023). https://doi.org/10.3390/polym15071785

34. Songsaeng, S., Thamyongkit, P., Poompradub, S.: Natural rubber/reduced-graphene oxide composite materials: morphological and oil adsorption properties for treatment of oil spills. J. Adv. Res. **20**, 79–89 (2019). https://doi.org/10.1016/j.jare.2019.05.007
35. Ababneh, H., Hameed, B.H.: Chitosan and chitosan composites for oil spills treatment: review of recent literature. J. Water Process Eng. **55**, 104193 (2023). https://doi.org/10.1016/j.jwpe.2023.104193
36. Olalekan, A.P., Dada, A.O., Adesina, O.A.: Review: silica aerogel as a viable absorbent for oil spill remediation. J. Encapsulation Adsorption Sci. **4**(4), 122–131 (2014). https://doi.org/10.4236/jeas.2014.44013
37. Wang, C., Lin, S.: Robust superhydrophobic/superoleophilic sponge for effective continuous absorption and expulsion of oil pollutants from water. ACS Appl. Mater. Interfaces **5**(18), 8861–8864 (2013). https://doi.org/10.1021/am403266v
38. National Research Council: Oil Spill Dispersants: Efficacy and Effects. The National Academies Press (2003)
39. Gong, C., et al.: Fast and all-weather cleanup of viscous crude-oil spills with Ti3C2TX MXene wrapped sponge. J. Mater. Chem. A **8**(38), 20162–20167 (2020). https://doi.org/10.1039/d0ta06480g
40. Singh, V., et al.: Novel natural sorbent for oil spill cleanup. Ind. Eng. Chem. Res. **53**(30), 11954–11961 (2014). https://doi.org/10.1021/ie5019436
41. Calabrese, L., et al.: Sorption capacity of polydimethylsiloxane foams filled with thermal-treated bentonite—polydimethylsiloxane composite foams for oil spill remediation. Materials **16**(13), 4818 (2023). https://doi.org/10.3390/ma16134818
42. Patalano, A., et al.: Scaling sorbent materials for real oil-sorbing applications and environmental disasters. MRS Energy Sustain. **6**(1) (2019). https://doi.org/10.1557/mre.2019.3
43. Zaarour, B., Liu, W.: Recent advances of textile sorbents for oil spills cleanup: a review. J. Ind. Text. **53** (2023). https://doi.org/10.1177/15280837231186652
44. Viju, S., Thilagavathi, G., Keertana, N.: Oil sorption properties of cotton comber noil/recycled polyester blended needle punched nonwoven fabrics—an optimization study. J. Ind. Text. **52** (2022). https://doi.org/10.1177/15280837221113090
45. Wei, Q., Oribayo, O., Feng, X., Rempel, G.L., Pan, Q.: Synthesis of polyurethane foams loaded with TiO_2 nanoparticles and their modification for enhanced performance in oil spill cleanup. Ind. Eng. Chem. Res. **57**(27), 8918–8926 (2018). https://doi.org/10.1021/acs.iecr.8b01037
46. Fu, C., Gu, L., Zeng, Z., Xue, Q.: One-step transformation of metal meshes to robust superhydrophobic and superoleophilic meshes for highly efficient oil spill cleanup and oil/water separation. ACS Appl. Mater. Interfaces **12**(1), 1850–1857 (2019). https://doi.org/10.1021/acsami.9b17052
47. Santos, O.S.H., Da Silva, M.C., Yoshida, M.I.: Synthesis and performance of different polyurethane foams as oil sorbents. J. Appl. Polym. Sci. **134**(41) (2017). https://doi.org/10.1002/app.45409
48. Doshi, B., Sillanpää, M., Kalliola, S.: A review of bio-based materials for oil spill treatment. Water Res. **135**, 262–277 (2018). https://doi.org/10.1016/j.watres.2018.02.034
49. Bhagwat, S.B., Jaspal, D.: Application of biodegradable materials in oil-spill cleanup: recent advances and future perspectives. Adv. Eng. Mater. **25**(20) (2023). https://doi.org/10.1002/adem.202300315
50. Nam, C., et al.: Practical oil spill recovery by a combination of polyolefin absorbent and mechanical skimmer. ACS Sustain. Chem. Eng. **6**(9), 12036–12045 (2018). https://doi.org/10.1021/acssuschemeng.8b02322
51. Owoseni, O., et al.: Release of surfactant cargo from interfacially-active halloysite clay nanotubes for oil spill remediation. Langmuir **30**(45), 13533–13541 (2014). https://doi.org/10.1021/la503687b
52. Wang, B., Karthikeyan, R., Lu, X., Xuan, J., Leung, M.K.H.: Hollow carbon fibers derived from natural cotton as effective sorbents for oil spill cleanup. Ind. Eng. Chem. Res. **52**(51), 18251–18261 (2013). https://doi.org/10.1021/ie402371n

53. Yadav, S., et al.: Recycling of thermoplastic polystyrene waste using citrus peel extract for oil spill remediation. J. Appl. Polym. Sci., **136**(33) (2019). https://doi.org/10.1002/app.47886
54. De Sousa Coelho, A.C.: Eco-insights on hydrocarbon bioremediation. Prog. Petrochem. Sci. **6**(2) (2024). https://doi.org/10.31031/pps.2024.06.000634
55. De Oliveira, A.F., et al.: Copper-contaminated substrate biosorption by Penicillium sp. isolated from kefir grains. Vide Leaf, Hyderabad eBooks (2023). https://doi.org/10.37247/pamicr3ed.3.23.14
56. Li, Z., et al.: Energy-optimized oil spill cleanup: joule-/solar-heating copper foam for efficient all-weather recovery of viscous crude oil. Ind. Eng. Chem. Res. **62**(33), 13133–13143 (2023). https://doi.org/10.1021/acs.iecr.3c01690

Chapter 2
Bio-based Foam Sorbents: Chemistry, Synthesis and Properties

Abstract This chapter delves into the synthesis and chemical properties of various bio-based foam sorbents, highlighting their potential for commercialization. It presents a detailed examination of the unique properties exhibited by these sorbents, discussing the innovative materials and processes used in their creation. The chapter is structured around two primary categories of bio-based foam sorbents: those derived from vegetable oils and those derived from biomass. Each section provides a comprehensive overview of the specific materials, synthesis methods, and the resulting properties of the foams, emphasizing their competitive advantages in the marketplace.

Keywords Castor oil · Polyol · Polyurethane · Vegetable oils · Coconut oil · Sunflower oil · Tannin · Polylactic acid · Cellulose · Polymerization · Cross linking · Biocomposite

2.1 Vegetable Oil-Derived Sorbent Foam

The increasing demand for sustainable materials has attracted the utilization of vegetable oil-derived polyols for PU foam production not only for their versatility but including their abundance, biodegradability, and low cost [1]. Vegetable oils are versatile because of their unique chemical structures which include multiple functional groups such as hydroxyls, double bonds, and esters. These functional groups facilitate the conversion of the vegetable oils into polyols that can react with isocyanates to form the PU foam sorbents which can be tailored according to the desired properties. These characteristics of vegetable oils make it viable to be an alternative to traditional petroleum-based polyols in the production of polyurethane foam sorbents [2, 3].

Recent advancements in the field showed the potential of vegetable oil-based PU foam sorbents for diverse applications, including oil spill cleanup and environmental remediation. These foams exhibited excellent sorption capacities due to their porous structure and hydrophobic properties which can be designed by selecting specific vegetable oils and modifying their chemical structures [4, 5]. For example,

A. A. Lubguban et al., *Bio-based Foam Sorbents*, Engineering Materials,
https://doi.org/10.1007/978-981-96-1460-8_2

PU foams synthesized from castor oil and other vegetable oils have demonstrated superior performance in their oil sorption capabilities, making them effective materials for removing oil from contaminated water [6]. The integration of vegetable oils in PU foam sorbents not only enhances the performance of these sorbents but also contributes to their biodegradability, aligning with the growing emphasis on sustainable materials in environmental remediations [7, 8]

Additionally, the shift towards bio-based PU foam sorbents moves closely with the global sustainability goals to reduce the dependence on fossil fuels and mitigate the environmental impact associated with the conventional PU foam sorbents production [9]. Renewable resources like vegetable oil are crucial in addressing the challenges posed by climate change and depletion of resources [10]. As research continues to explore the potential of different vegetable oils including castor, coconut, and soybean oil, the scope for developing innovative PU foam sorbents that are both high-performing and environmentally friendly is expanding [11, 12].

2.1.1 Castor Oil-based Polyurethane Foams

Castor oil, derived from the seeds of the Ricinus communis plant, has garnered significant attention in the field of sustainable and green innovations. It has been used in developing bioplastics, bio-composites, adhesives, coatings, bio-based elastomers, and particularly in PU foams synthesis. Its chemical composition characterized by high content of hydroxyl groups allows it to effectively function as a renewable polyol in the production PU foams, replacing the traditionally derived PU foam from petroleum-based sources. Castor oil is primarily made up specifically of ricinoleic acid (approximately 90%) shown in Fig. 2.1, is an important composition for PU synthesis [13], it is a combination of hydroxyl and unsaturated fatty acids that contributes to the formation of urethane linkages which make up polymer networks [7]. Moreover, castor oil's hydroxyl number which is approximately 2.7 per triglyceride allows greater degree of crosslinking in the polymer network resulting in improved mechanical properties and foam stability [14].

The versatility of castor oil in PU foam synthesis is thoroughly recorded, for instance, it has been utilized to successfully develop flexible PU foam with different compositions, which demonstrates that castor oil can significantly influence the mechanical properties and thermal stability of the resulting materials [15]. Research indicates that the incorporation of castor oil in PU foam formulations can enhance the compressive strength and improve the overall performance of the foams, making them suitable for thermal insulation and protective equipment applications [16]. Moreover, castor oil's inherent properties allow it to be used without requiring the modification of introducing hydroxyl group in the parent chain as it inherently contains hydroxyl group in its parent chain compared to most vegetable oils and plant derived polyols [17].

Other than flexible foams, castor oil has also been employed in the production of rigid polyurethane foam. Studies have shown that it can be modified with

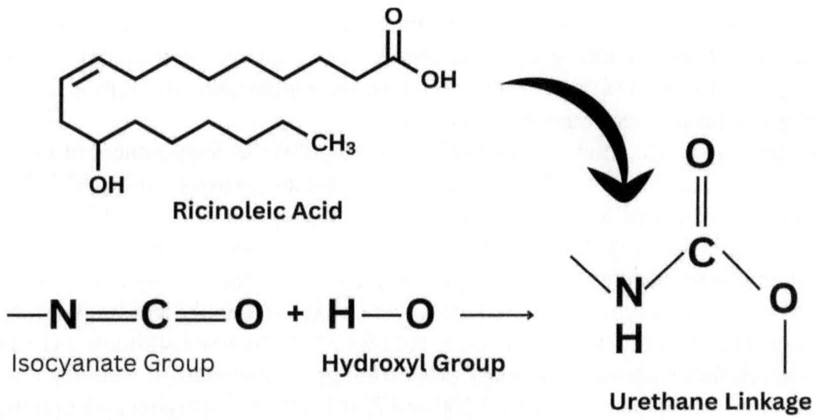

Fig. 2.1 Important components in polyurethane synthesis

maleic anhydride to produce biodegradable foams, expanding its application in environmentally-friendly materials [18]. The ability to synthesize both flexible and rigid foams from castor oil highlights its adaptability and potential to replace conventional petrochemical-based polyols across various sectors, including construction and automotive industries [19]. Furthermore, the use of castor oil in PU formulations has been linked to the development of composite materials. For example, castor oil-based PU have been incorporated with natural fillers such as bamboo charcoal to create low-density composites that maintain desirable mechanical while being more sustainable [20]. This integration of castor oil with other materials illustrates its role in enhancing the functionality and eco-friendliness of PU products.

The synthesis of castor oil-based PU foam involves a systematic approach that integrates the properties of castor oil with traditional PU chemistry. Initially, the process begins with choosing the raw materials such as the polyol component, where in this case is the castor oil-based polyol, and the diisocyanate component, often toluene diisocyanate (TDI) or methylene diphenyl diisocyanate (MDI), blowing agent, fillers, and additives. The materials are then combined in a controlled manner to achieve the desired NCO/OH ratio which is crucial for determining the final properties of the PU foam [21]. The synthesis follows a two-step polymerization approach where in the first step, the polyol reacts with diisocyanate to form a prepolymer. This reaction can be influenced by several factors, including temperature and mixing speed, which affects the molecular weight and characteristics of the resulting polymer [22]. After the formation of the prepolymer, chain extenders such as glycerol or 2,2-dimethylol butanoic acid (DMBA) may be introduced to enhance mechanical properties and crosslinking density, which ultimately influences the flexibility and strength of the final product [6]. After the polymerization process is the curing process which can vary with options like thermal curing or moisture curing, where ambient moisture reacts with isocyanate groups to form stable urethane linkages. The curing conditions such as temperature and humidity are critical in determining the final properties of the PU foams such as elasticity, hardness, and thermal stability [23]. Moreover, the incorporation of co-monomers or additives, such as polydimethylsiloxane polyol (PMDS) can further enhance the mechanical strength and modify the physical properties of the PU foam [24].

A significant utilization of castor oil-based polyol is the development of PU foam sorbent intended for oil spill remediation. For instance, castor oil-based PU foam sorbent is synthesized to test its efficiency for oil sorption of diesel oils S10 AND S500. The PU foam sorbents were synthesized by mass mixing castor oil-based polyol and diisocyanate (4,4-diphenylmethane diisocyanate) using different molar masses of the prepolymer. The materials were mixed manually and transferred to a cylindrical plastic mold then were left to cure for 24 h. The foam sorbents were tested and showed that with the same mass proportion of the reactants, it exhibits a higher diesel sorption efficiency (5 g/g for S500 and 7 g/g for S10) and higher hydrophobicity (110.3°) due to its higher pore diameter and lower density. These characteristics are accounted for by the mass proportion of the reactants which makes sure that all reactants are consumed. The maximum sorption capability of these foams occurred between 25 to 40 h of the sorption test. It was noted that the viscosity of the diesel influenced the sorption process as the foam sorbents performed better with S10 diesel than diesel S500 due to its higher viscosity [25].

Another notable innovation is the synthesis of castor oil-based PU foam sorbent grafted with octadecyltrichlorosilane (C18) functionalized diatomaceous earth (DE) particles as a surface modifying agent to enhance the oil adsorption capacity and hydrophobicity of the PU foam. The C18-DE modification in Fig. 2.2 was done by mixing the C18 and unmodified DE particles with hexane as solvent. The mixture was heated at 50 °C for 5 h and the solid particles from the solution were separated by centrifugation, followed by washing of the solid particles using hexane to separate the unreacted C18 from C18-DE modified particles. The modified particles were then air dried in the oven at 70 °C for 2 days [26].

The use of castor oil polyol for PU foam sorbent for oil spill cleanup necessitates surface modification due to its inherent limitations when used in its unmodified state. Castor oil polyols generally exhibit lower mechanical strength compared to petroleum-based polyols and often suffer from reduced tensile strength and elongation at break [16]. The lower crosslinking density of castor oil polyols is due to their low hydroxyl number (approximately 160 mg KOH/g) that leads to softer foams that may not withstand mechanical stress effectively [27]. Research also shows that higher castor oil content leads to lower degradation temperature and faster degradation rate which affects its thermal stability and compromises their performance in high temperature environments [28]. The foam sorbents from castor oil may not exhibit optimal hydrophobicity and oleophilicity which can lead to subpar oil sorption capacities and may not effectively repel water while attracting oil. Moreover, the low functionality of the oil can limit the crosslinking during the polymerization process leading to inadequate structural stability and performance [29]. Additionally, chemical compatibility is also considered as castor oil may not be compatible with other additives and fillers that may improve its properties. Surface modification enhances these limitations and therefore improves its performance [30].

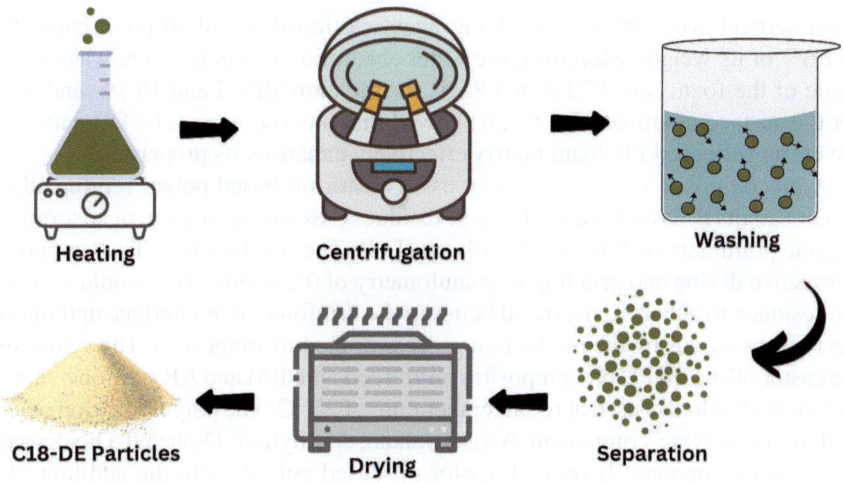

Fig. 2.2 Modifying diatomaceous earth (DE) particles with octadecyltrichlorosilane (C18)

The role of C18 in the PU foam synthesis is it serves as a silane coupling agent that modifies the surface characteristics of DE, which is a naturally occurring, porous material composed primarily of silica. The incorporation of C18 into the synthesis process improves the hydrophobicity of the DE, thereby enhancing the sorption capacity of the PU foams for oily pollutants, which is already demonstrated in studies focusing on the functionalization of DE with alkyl silanes [26]. Moreover, DE contributes significantly to the mechanical and thermal properties of PU foams. Its porous structure not only provides a lightweight filler but also increases the surface area available for interaction with contaminants, which is crucial for applications in environmental remediation. The modification of DE with C18 further enhances these properties by creating a hydrophobic surface that facilitates the absorption of non-polar organic solvents, making the biobased PU foam more effective in oil spill clean-up scenarios [31]. Moreover, the use of biobased polyols, such as those derived from castor oil, in conjunction with C18-modified DE, allows for the development of eco-friendly PU foams. These foams exhibit favorable mechanical properties and thermal stability, which are essential for their performance as sorbents [32]. The combination of biobased materials with functionalized DE not only promotes sustainability but also enhances the overall effectiveness of the foam in various applications, including environmental cleanup and as a lightweight structural material [33].

The synthesis of the castor oil-based PU foam sorbent modified with C18-DE particles is represented in Fig. 2.3 where castor oil polyol, blowing agent, and surface modifying agent is mixed, then followed with the addition of isocyanate. The mixture is stirred and transferred to a test tube and left to cure for 2 days. The researchers compared the foam with surface modifying agent and the foam sorbent without modification shown in Fig. 2.4, the result showed that the foam sorbent without modification exhibited an oil sorption capacity of 45.2% of its weight, while the foam sorbent with surface modifying agent exhibited an oil sorption capacity of 131.6% of its weight. Moreover, the water absorption percentage and water contact angle of the foams are 57.5% and 85.5° for the unmodified and 16.7% and 129.8° for the surface modified. The result showed that incorporating C18-DE particles in the castor oil-based PU foam matrix effectively enhances its properties.

A biocomposite is also developed using castor oil-based polyol reinforced with bagasse malt (BM) and acerola bagasse residues (AR) for separation or adsorption of organic pollutants such as oil and solvents [34]. The residues were first prepared by convective drying and grinding to granulometry of 0.250 mm. This would ensure for the residues to adhere and expand better in the reinforcement interface and improve the uniformity during the production process of the biocomposites. The synthesis of the castor oil-based PU biocomposites reinforced with BM and AR is by one-shot free expansion method is used at room temperature of 25 °C. The polymerization reaction is done by mixing component A (isocyanate, Methylene Diphenyl Diisocyanate-MDI) and component B (polyol, castor oil-based polyol) with the addition of the

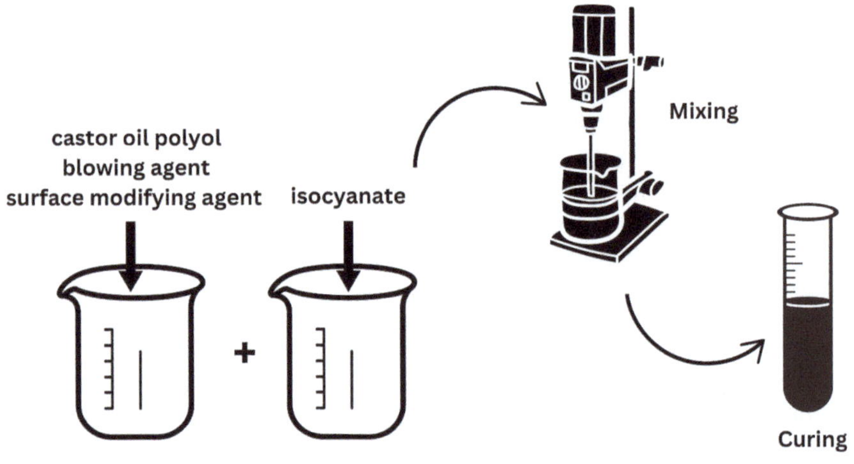

Fig. 2.3 Synthesis of the castor oil-based PU foam sorbent modified with C18-DE particles

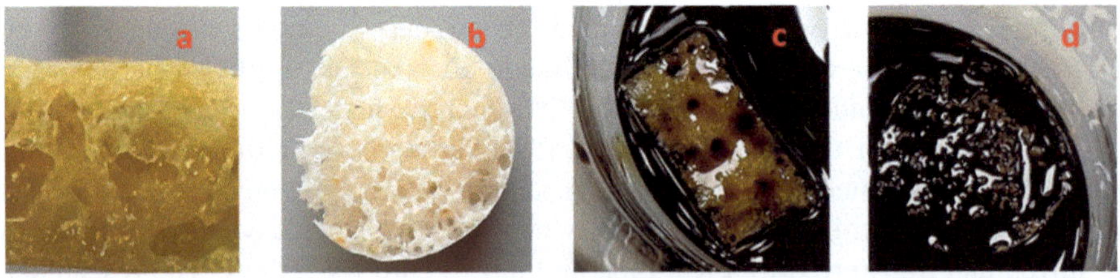

Fig. 2.4 Oil absorption studies in bio-based polyurethane (PU) foams: **a** unmodified PU, **b** PU with C18-DE modification (PU-C18-DE), after 20 min of contact with crude oil, and **c** unmodified PU and **d** PU-C18-DE foam conditions post-absorption. Adapted from Perera et al. [31], licensed under CC BY 4.0. Available at: https://creativecommons.org/licenses/by/4.0/

residues. The residues were added individually in proportions along with components A and B, and was stirred using a mechanical stirrer and was left for 30 s cream time and was cured for 24 h.

The incorporation of bagasse residues as a biocomposite in PU foam sorbents is due to the reason that bagasse is rich in lignocellulosic materials that can contribute to attain high surface area and porosity of the PU foam. These characteristics are important for adsorption applications as they are responsible for trapping the oil and organic solvents within the PU foam matrix. The porous structure of bagasse can enhance the interaction between oil and organic solvents (adsorbate) and the biocomposite foam (adsorbent) [35]. Additionally, the functional groups present in the lignocellulosic materials can enhance the chemical interaction with different organic compounds, improving the adsorption efficiency [36], it also alters the chemical composition of the PU foam as the bagasse participates in the urethane interaction due to the presence of hydroxyl group in its structure which reacts with isocyanate that leads to enhanced overall reactivity of the biocomposite during PU synthesis [37].

The hydroxyl groups present in the bagasse can also participate in the gelling reaction that affects the blowing reaction during synthesis which could reduce cell size [38], moreover, it also acts as nucleation sites, enhancing the formation of bubbles and contributing to the overall structure of the foam [37], making it suitable for various applications. The fibrous nature of bagasse provides reinforcement within the foam structure, which can lead to better load-bearing capabilities and durability, essential for oil spill cleanup and other industrial applications [39].

The properties of the foams showed that biocomposites with 20% residue showed better performance compared to other biocomposites with different proportions. The sorbents were confirmed to adsorb motor oil effectively with 0° contact angle and repel water efficiently with $133.2 \pm 0.2°$ and $144.4 \pm 0.2°$. This is attributed to the incorporation of residues in the PU foam sorbents. The increase in contact angle with the biocomposite foams in comparison with the foams without residues affirms the successful adhesion of the residues in the polymer matrix, resulting in void reduction. The increase in hydrophobicity resulted from the increase in roughness which favors hydrophobic characteristics [40]. These characteristics of foam considers them as hydrophobic materials which is an important property for foam sorbents intended for oil spill cleanup. The adsorption capacity of the foams were 1–33 times their weight, it increases as the percentage of residues increases. The biocomposites with AR showed better adsorption capacity for oil of $827.4 \pm 98.1\%$ while with BM it has $340.7 \pm 40.4\%$, additionally, biocomposite with AR showed 262% higher adsorption for kerosene in comparison to biocomposite with BM. These composites can be recovered and recycled for 10 times with AR composites having higher recovery index than BM composites which is attributed to the larger pore structure of AR composites that makes the release of oil easier. Moreover, the composites were tested for continuous oil/water separation through a suction system set-up in aqueous medium and the biocomposites performed well with the oil penetrating the biocomposites and decompressing within 20.5 ± 0.9 s and 39.5 ± 1.9 s for AR and BM composites, respectively. However, exposure to saline conditions decreases the adsorption capacity of the biocomposite which calls for enhancement and modification to improve the performance of the biocomposites in saline environments. Overall, the biocomposites showed great potential for oil spill cleanup and recovery without bringing harm to the environment.

Another biocomposite using castor oil-based polyol reinforced with açaí seed residue is developed for the adsorption of organic pollutants such as waste cooking oil (WCO). The açaí seeds, lignocellulosic biomass, were collected, ground, and sieved to a granulometry of 28 mesh. This preparation enhanced the interface adhesion between the castor oil-based polyurethane (PU) and the açaí seed residue, improving uniformity in the production process of the biocomposite. The synthesis involved the one-shot expansion method, where castor oil-based polyol and isocyanate (MDI) were mixed in a 1:1 ratio and reacted at room temperature. The polymerization occurred at 55 °C, and the foam was cured for 24 h. The incorporation of the açaí seed fibers into the polyurethane matrix aimed to improve the biocomposite's mechanical and sorption properties.

The use of açaí seed fibers in the polyurethane foam is based from the fact that açaí seed fibers is rich in cellulose, hemicellulose, and lignin, which increases the surface area and porosity of the biocomposite, enhancing its capability to adsorb oils. This porous structure, essential for adsorption, enables efficient trapping of oil within the foam matrix. The functional groups present in lignocellulosic materials facilitate interactions with organic compounds, improving the overall adsorption efficiency. The hydroxyl groups in the açaí seed fibers contribute to urethane linkages during the PU synthesis, enhancing the chemical reactivity and compatibility of the composite.

The biocomposite showed significant improvements in hydrophobicity and adsorption capacity compared to neat polyurethane. The water contact angle increased due to the presence of the açaí fibers, which raised the surface roughness and hydrophobic characteristics, critical for oil sorption applications. The foam achieved an oil adsorption capacity of up to 10.42 g of WCO per gram of sorbent, demonstrating higher sorption efficiency in a short time. These findings confirm that castor oil-based PU reinforced with açaí seed residue is a promising eco-friendly material for oil spill cleanup. Additionally, the biocomposite could be reused in multiple cycles with a retention of approximately 26% of its sorption efficiency after seven cycles.

Another castor oil-based biocomposite foam reinforced with coffee husk residue (CR) was developed for the adsorption and removal of diesel oil (S10 and S500) from contaminated water. The coffee husks, a lignocellulosic byproduct from coffee processing, were ground to a fine particle size (28 mesh) and incorporated into the polyurethane (PU) matrix at different concentrations (5%, 10%, 20%, 30%, and 40% w/w). The synthesis of the biocomposite was carried out using a one-shot expansion method at room temperature, where castor oil-based polyol and diisocyanate (MDI) were mixed with the coffee husk residue. This process resulted in a porous PU composite foam with improved surface roughness, hydrophobicity, and mechanical properties, making it highly efficient for oil sorption applications.

The inclusion of coffee husk residues into the PU foam significantly altered the foam's pore structure, shifting from an open-cell to a partially closed-cell configuration, which enhanced the oil sorption capacity. Coffee husks, rich in cellulose and lignin, contributed to increasing the foam's surface area and hydrophobicity, important factors for oil adsorption. The hydrophobic nature of the foam, evidenced by the increase in contact angle, facilitated selective oil uptake over water, a critical feature for oil spill remediation. The FTIR analysis confirmed the successful interaction between the coffee husk and polyurethane matrix, indicating strong bonding that enhanced the material's sorption performance.

The biocomposite foam demonstrated high efficiency in diesel oil removal, with the PU + 40% CR composite absorbing 5.7 g of Diesel S10 and 6.2 g of Diesel S500 per gram of foam within 1 h. These sorption capacities are higher than those observed in neat polyurethane foams. Additionally, the foam exhibited excellent reusability, maintaining over 80% of its sorption efficiency after 14 cycles, making it suitable for repeated use in oil spill cleanup operations. This makes castor oil-based polyurethane foams reinforced with coffee husks a promising eco-friendly material for environmental remediation.

Another innovation developed using castor oil-based polyol is the superhydrophobic polyurethane foam (PU) with lignin, and silicon carbide (SiC) nanoparticles for efficient and recyclable oil–water separation. In the study, SiC nanoparticles were first treated with 1H, 1H, 2H, 2H-perfluorododecyltrichlorosilane (FDTS) to make them superhydrophobic. These modified F-SiC nanoparticles, along with liquefied lignin, were incorporated into a castor oil-based polyurethane matrix via a one-step foaming process. The foam's structure consisted of a three-dimensional interconnected network, which is crucial for oil absorption. The presence of F-SiC increased the foam's water contact angle to 151.7°, confirming its superhydrophobicity and making it highly efficient for oil–water separation applications.

The combination of castor oil and lignin in the PU foam significantly enhanced its eco-friendliness, as both materials are bio-based and biodegradable. Mechanical tests showed that the foam maintained its structural integrity, even after 100 compression cycles, with minimal plastic deformation, which is critical for repeated use in oil spill cleanup. The addition of F-SiC improved the foam's photothermal properties, with surface temperatures reaching up to 88.8 °C under 1 kW/m^2 of sunlight. This photothermal capability reduces the viscosity of crude oil, making it easier for the foam to absorb the oil efficiently.

The foam exhibited high oil adsorption capacities, absorbing 3–17 times its own weight in various oils and organic solvents, depending on the density of the liquid. Even after 20 cycles of adsorption and desorption, the foam retained over 95% of its original adsorption efficiency. Additionally, the foam showed excellent self-cleaning properties, with water droplets removing dirt from the surface, further enhancing its practical applications. The foam also degraded in an alkaline solution, proving its environmental compatibility and biodegradability.

An eco-friendly sorbents is also developed by incorporating macadamia nutshell waste (MW) into castor oil-based polyurethane (PU) foam for the removal of crude oil and diesel from water. The macadamia nutshell waste, a lignocellulosic byproduct, was prepared by drying, crushing, and sieving to a size of 28 mesh. The macadamia waste was mixed with castor oil-based polyol and a prepolymer (MDI) to create biocomposites with varying MW contents (5%, 10%, 15%, and 20% by weight) using a one-shot foaming method. The biocomposites exhibited a porous structure that is crucial for oil sorption, with increasing hydrophobicity and surface roughness proportional to the macadamia content. The foam's contact angle increased to 119.1° for the sample with 20% MW, confirming its hydrophobic nature and suitability for oil–water separation.

The addition of macadamia nutshell waste into the PU matrix altered the foam's morphology, reducing pore size and increasing the number of pores per area. This structural modification enhanced the foam's ability to absorb oils. Maximum oil absorption capacities of 7.3 g/g for crude oil and 3.9 g/g for diesel were achieved with the PU + 20% MW biocomposite. The macadamia waste acted as nucleation sites during the foam's expansion process, contributing to smaller cell sizes and improved mechanical properties. These structural changes enabled the biocomposites to outperform neat PU foams, especially for heavy oil sorption.

The biocomposites also demonstrated excellent reusability, maintaining high sorption capacities over 30 sorption–desorption cycles for crude oil, with a slight decline in efficiency for diesel. This reusability, along with the foam's biodegradability and cost-effectiveness, makes it an attractive solution for large-scale oil spill remediation.

Polyurethane (PU) foams reinforced with Australian palm fiber residues (PR) were also synthesized to improve the oil sorption capacity, focusing on the removal of crude oil and S500 diesel from seawater. The polyurethane matrix was synthesized using castor oil-based polyol and isocyanate in a 1:1 ratio, combined with varying concentrations of palm fiber residues (5, 10, 15, and 20% by weight). The palm residues were processed into two particle sizes (28 and 35 mesh), influencing the foam's structure. The addition of palm fibers decreased pore sizes and increased the homogeneity of the open-cell structure in the foams, which is crucial for oil sorption. The use of PR led to significant improvements in sorption capacity, with foams exhibiting a maximum absorption capability of 24 g/g for crude oil and 18 g/g for diesel S500, far surpassing the capabilities of pristine polyurethane foam.

The mechanical properties of the foam were enhanced by the palm fiber content, with the 20% PR composite showing the best results in terms of oil absorption and structural integrity. The porous structure, combined with the hydrophobic characteristics of the palm fibers, played a critical role in the foam's efficiency in separating oil from water. Life Cycle Assessment (LCA) revealed that the incorporation of palm fibers reduced the environmental impact of the foam, particularly when compared to conventional residue disposal methods such as burning. The LCA analysis emphasized that using 20% PR provided the most sustainable option, with lower climate change and human toxicity impacts.

The foams showed excellent reusability, maintaining more than 50% of their adsorption efficiency after 16 sorption/desorption cycles for crude oil and diesel. This makes the PU-20% PR composites particularly promising for long-term oil spill cleanup operations. In oil/water systems, especially in seawater, the foam's performance slightly decreased due to salinity, but it still retained superior sorption capabilities compared to conventional foams. The Langmuir adsorption model predicted maximum adsorption capacities of 30.39 g/g for crude oil and 20.79 g/g for diesel in seawater.

2.1.2 Sunflower Oil-Based Polyurethane Foams

Sunflower oil has emerged as a significant component in sustainable materials and green chemistry. It is derived from the seed of Helianthus annuus plant with a fatty acid composition consisting of linoleic acid that allows it to be an effective bio-based polyol in the synthesis of PU foam synthesis in replacement to petroleum-derived polyols. The high degree of unsaturation in sunflower oil facilitates the formation of the urethane linkages, which are crucial for developing robust polymer networks in PU foam [41]. Furthermore, the inherent properties of sunflower such as its viscosity and thermal stability contributes to the overall performance of PU formulations, enhancing their properties and capabilities [42].

Research has demonstrated that sunflower oil can be utilized to create both flexible and rigid PU foams, showcasing its versatility across various applications, including construction and automotive industries [41, 42]. The incorporation of sunflower oil into PU formulations has been linked to improved compressive strength and durability, making these materials suitable for demanding applications [41]. Additionally, modifications such as maleation of sunflower oil can lead to the production of biodegradable foams, further expanding its utility in environmentally friendly materials [42]. The adaptability of sunflower oil in PU synthesis is further evidenced by its successful integration with natural fillers, resulting in composite materials that maintain desirable mechanical properties while promoting sustainability [43].

Sunflower oil serves as an optimal precursor for polyol synthesis through a one-step hydroxylation reaction. These double bonds are converted into hydroxyl groups, a key functional group for polyurethane production. This transformation is achieved using hydrogen peroxide (H_2O_2) and formic acid (HCOOH) as oxidizing agents, which yield a bio-polyol with a hydroxyl value of approximately 182 mg KOH/g and around 3.5 OH groups per molecule, essential for crosslinking within the PU matrix. The chemical changes are validated through nuclear magnetic resonance (NMR) spectroscopy, where olefinic proton signals characteristic of sunflower oil disappear, and new signals representing hydroxyl groups appear. Fourier-transform infrared (FTIR) analysis further confirms hydroxylation, showing the emergence of broad hydroxyl stretching bands near 3496 cm^{-1}, indicating successful conversion to polyol [44].

The composite synthesis combines the bio-polyol with petrochemical polyols in a 20:80 ratio, striking a balance between renewability and mechanical robustness. To further enhance the material's oil adsorption capabilities, sugarcane bagasse, a lignocellulosic agricultural byproduct, is incorporated as a filler. The bagasse is initially processed by size reduction (to approximately 1 mm) and hot water treatment at 170 °C, which partially removes hemicellulose and exposes cellulose fibers, increasing the filler's surface area and hydrophobicity. This pretreatment enhances compatibility within the PU matrix, facilitating a stable, porous structure that maximizes oil uptake.

During synthesis, the bio-polyol and petro-polyol blend is combined with the sugarcane bagasse particles, followed by the addition of diphenylmethane diisocyanate (p-MDI) as the isocyanate component. The reaction mixture is poured into a mold where the curing reaction forms a foamed PU structure with embedded bagasse particles. This process creates a porous, open-cell architecture, crucial for efficient adsorption of oils by providing ample pathways for oil to penetrate the material's internal structure [44].

The addition of sugarcane bagasse as a filler within the polyurethane matrix serves several essential purposes. Sugarcane bagasse, a lignocellulosic byproduct of the sugar industry, is rich in cellulose, hemicellulose, and lignin, making it highly porous and hydrophobic once processed. By incorporating this biomass as a filler, the composite's porosity and overall surface area are significantly enhanced, which are critical properties for materials used in oil adsorption. The bagasse filler is pre-treated with hot water at 170 °C to remove some hemicelluloses and increase the cellulose's hydrophobicity, rendering it more compatible with the PU matrix and maximizing its adsorption efficiency. This filler not only increases the surface area for oil interaction but also reinforces the structural integrity of the open-cell PU, creating a composite material with interconnected channels conducive to high-capacity oil absorption. The bagasse's natural oleophilicity (affinity for oil) complements the hydrophobic PU matrix, improving the composite's selectivity for oil over water and enhancing overall adsorption performance [44].

The SC-PU composite's high porosity and hydrophobicity make it an efficient material for oil spill adsorption. In tests, this bio-composite achieves an adsorption capacity of up to 15.2 g of oil per gram of sorbent, a notable improvement over conventional PU foams without biomass fillers. This performance is attributed to the synergy between the open-cell PU matrix and the porous sugarcane bagasse, which enables oil to migrate efficiently through the interconnected channels of the composite material. Additionally, the bio-polyol's hydrophobicity and the bagasse's oleophilic nature enhance oil selectivity and retention, making the SC-PU composite a practical choice for oil spill remediation. Compared to synthetic alternatives, the SC-PU offers added benefits of biodegradability and reduced environmental impact, furthering its appeal as a sustainable option for large-scale environmental cleanup.

In practical terms, this bio-based composite can be deployed in oil spill scenarios, where its buoyancy and high oil-to-water adsorption selectivity make it particularly efficient. Moreover, its biodegradable nature addresses concerns of secondary pollution commonly associated with synthetic materials, positioning the SC-PU composite as a sustainable alternative for large-scale environmental cleanup efforts.

2.1.3 Palm Olein-Based PU Foams

Palm olein has been identified as an effective bio-based polyol for the synthesis of polyurethane (PU) foams, providing a sustainable alternative to petroleum-based reagents. The synthesis of this polyol from palm olein involves a series of chemical reactions, including transesterification and epoxidation, which result in a polyol characterized by functional hydroxyl groups [45]. These hydroxyl groups are crucial for their reactivity with isocyanates, leading to the formation of urethane linkages essential for creating polyurethane foam [46]. The palm olein-based polyol was incorporated into the foam at varying concentrations (25, 50, and 75%), and the resulting material exhibited a porous structure with interconnected cells, enhancing its ability to serve as a sorbent material. The interconnected pores allow for increased oil sorption, making this foam particularly suitable for environmental applications such as oil spill remediation.

The mechanical properties of the palm olein-based polyurethane foam were also evaluated. While the introduction of palm olein polyol led to a slight reduction in tensile strength and Young's modulus compared to commercial polyurethane, it significantly increased the elongation at break. This suggests improved flexibility, making the foam more adaptable for applications requiring material stretch. Additionally, the thermal stability of the foam was analyzed using thermogravimetric analysis (TGA) and differential scanning calorimetry (DSC). Although the incorporation of palm olein polyol decreased the thermal stability of the foam, this trade-off was balanced by its enhanced flexibility and sorption properties, which are crucial for practical environmental uses.

The palm olein-based polyurethane foam demonstrated impressive oil and dye sorption capacities, making it a strong candidate for environmental remediation. The foam exhibited higher oil sorption capacities, absorbing 2–4 times more oil than commercial polyurethane foams [46]. This was attributed to the interconnected pore structure, which facilitated the efficient absorption of oil [47]. Additionally, the foam showed a capacity for dye sorption, particularly for methyl orange, an anionic dye, further expanding its potential application in removing water pollutants. The combination of its porous structure, hydrophilic nature, and flexibility positions palm olein-based polyurethane foam as an eco-friendly and effective sorbent material for both oil and dye pollutants.

2.1.4 Coconut Oil-Based PU Foam

Coconut oil, a renewable and abundant resource, has garnered attention in recent years for its application in the development of polyurethane (PU) foams. This interest is largely attributed to the inherent superoleophilic and hydrophobic properties of coconut oil, which render it particularly suitable for oil spill remediation efforts. The transformation of coconut oil into a polyol—a critical component in the production of polyurethane foam—forms the basis of this innovative approach [48].

The synthesis of coconut oil-based foam sorbents primarily involves a two-step process that begins with glycerolysis. In this process, the triglycerides present in coconut oil are broken down into monoglycerides. The resulting coconut oil-based polyol is subsequently mixed with petroleum-based polyols, surfactants, and other additives to create a polyurethane foam characterized by a significant bio-based content. This foam is produced using the conventional one-shot foaming method, wherein a polyisocyanate is combined with the polyol blend. The resulting foam is distinguished by its high porosity and rough surface texture, both of which contribute to its exceptional oil sorption capacities.

Coconut oil exhibits unique oleophilic and hydrophobic properties, primarily due to its molecular structure and composition. The predominant component of coconut oil, lauric acid, features a long hydrophobic carbon chain that enhances the oil's oleophilicity. This characteristic enables coconut oil to effectively attract and absorb oil-based substances while simultaneously repelling water, a fundamental requirement for effective oil spill cleanup materials. The hydrophobic nature of coconut oil is further accentuated by its relatively low polarity compared to water, resulting in a high contact angle when in contact with water, thereby promoting water repellency [49].

The oleophilic behavior of coconut oil is also attributed to its capacity to form a complex network of interactions with oil molecules. The long-chain fatty acids present in coconut oil facilitate strong van der Waals forces with oil, thereby enhancing its affinity for oil over water. This property is crucial for the development of sorbent materials that can selectively absorb oils while maintaining structural integrity in aqueous environments. Moreover, the presence of functional groups in the fatty acid chains can lead to increased surface roughness when coconut oil is incorporated into polyurethane foams, further enhancing the oleophilic and hydrophobic characteristics of the resulting material [49].

The incorporation of coconut oil as a polyol in the PU matrix significantly enhanced the mechanical properties of the foam. The tensile strength increased by 47% compared to conventional petroleum-based PU foams, making it durable for industrial use. Moreover, thermogravimetric analysis (TGA) demonstrated that the foam had superior thermal stability, largely due to the saturated nature of coconut oil's fatty acid chains. These thermal properties make the foam well-suited for environments requiring resistance to high temperatures. Additionally, the foam's open-cell structure, as seen in SEM analysis, facilitated a high degree of oil retention, enabling it to absorb between 14.89 and 24.65 times its weight in oil. This capacity for high-volume absorption, paired with its mechanical robustness, positions the coconut oil-based foam as an ideal candidate for addressing oil spills and industrial wastewater treatment [49].

Coconut oil-based PU foam also showed excellent recyclability, maintaining its oil absorption efficiency after 20 cycles of use with minimal degradation in performance. This reusability is crucial for practical applications, particularly in large-scale oil spill remediation, where cost-efficiency and material longevity are essential. Additionally, the foam exhibited strong chemical stability, resisting degradation in acidic, saline, and basic environments [49]. This durability across various harsh conditions further emphasizes its potential for deployment in diverse environmental cleanup operations. The use of coconut oil in PU foam synthesis thus represents a sustainable, environmentally friendly alternative to petroleum-based products, providing a solution that balances performance, reusability, and ecological impact.

2.2 Biomass-Derived Sorbent Foams

The growing need to manage waste generated by human activities has led to a heightened interest in sorbent foams, especially for environmental cleanup. Oil spills, although often unintended, have a serious impact on marine and coastal ecosystems and even pose risks to human health. A promising solution to these types of misfortunes are sorbent foams, although there are other types of sorbents, the potentials of foams stand out due to their high absorption capacity, lightweight nature, and ease of deployment across contaminated areas. However, the current method of producing sorbent foams (or foams in general) are unsustainable due to their reliance on petroleum-based chemicals and energy-intensive manufacturing processes, which are limited resources or economically impractical methods for production. Biomass-based foam sorbents have gained traction due to their renewable raw materials biodegradability, and availability. Unlike traditional foams, which often use nonrenewable resources and generate significant waste, these biomass-based alternatives utilize natural materials like agricultural by-products or plant-derived polymers.

The main advantage of biomass-derived sorbent foams compared to their renewable counterparts is that both the source and materials are sustainable. While vegetable oils may be renewable, their usage in foam production can lead to land use changes and competition in food production. On the other hand, biomass-derived sorbent foams, such as those made from agricultural or industrial waste, reduce competition with food production since they utilize resources that might otherwise be discarded and would contribute to waste management issues. The utilization of biomass supports a circular economy model wherein waste materials are used and converted into high-value products, especially since the world today is facing a problem of oil spills, which are detrimental to the environment.

2.2.1 Lignin-Based PU Foams

Lignin, a polymer present in plant cell walls and is the earth's second most abundant renewable polymer, is a complex, 3D aromatic polymer that is primarily composed of three types of monolignols: p-coumaryl, coniferyl, and sinapyl alcohols, which are linked through various types of bonds to form a heterogeneous network [50]. It constitutes 15–40% of lignocellulosic biomass, making it the largest non-carbohydrate component and is the reason to the hydrophobic properties of plant cell walls [51]. Lignin has been widely used for many applications; for example, lignin can undergo thermochemical and catalytic conversion for it to be used as a biomass-based fuels for combustion, which can reduce our dependency on fossil fuels and reduce our carbon footprint [52].

Recently, lignin has been explored as the raw material in the production of bio-based polyols, which are the building blocks of polyurethanes. Polyurethanes have been known for their effectiveness in various applications, such as cushioning and comfort, but now they are also promising for oil spill remediation. Their porous structure and surface area facilitate the sorption process, while their reusability makes them more sustainable in the long run. However, the sorption capacity of PU is highly dependent on the nature of its materials and lignin is a good candidate for natural sorbents. Lignin's chemical structure has fewer hydroxyl groups than cellulose's, mainly composed of aromatic compounds. This makes it less hydrophilic and more hydrophobic than cellulose [53].

Lignin can be classified into different types, which are usually based on its source, extraction method, and chemical structure. Table 2.1 summarizes the different types of lignin, their characteristics, and their applications in polyol production.

The process of making polyurethane sorbents for oil spills from lignin usually employs three steps: first is the liquefaction of lignin, second is the foaming process, and last is the surface modification of the produced polyurethane foams. In the context of sorbents for oil spills, they usually use enzymatic hydrolysis lignin or enzymatic lignin due to its lesser-modified nature, which preserves its inherent hydrophobic properties suitable for sorbent foams [61–64].

First, the lignin is dissolved in a solvent system comprising polyethylene glycol and glycerol. The chain extender is PEG and glycerol introduces hydroxyl groups into the polymer. Then sulfuric acid is added, which catalyses the reaction and de polymerizes the lignin. It breaks the ether bonds in the lignin and releases more hydroxyl groups. This process is kept steady for an hour at 140 °C under a nitrogen atmosphere to prevent oxidation. Finally, the liquefied mixture is neutralized with sodium hydroxide, and then water is removed from the excess quantity of the mixture using a rotary evaporator.

After liquefaction, the lignin-based polyols are reacted with isocyanates to form polyurethane foams. The commonly used isocyanate in this process is hexamethylene diisocyanate (HDI), which reacts with the hydroxyl groups in the lignin polyol to form urethane linkages. The reaction is balanced by maintaining a 1:1 [NCO]/[OH] ratio, ensuring that all available hydroxyl groups participate in crosslinking, which results in a strong and stable foam. Although depending on the properties required, other types of isocyanates are used. But in the context of sorbents, HDI is used due to its relatively hydrophobic nature when reacted with the polyol for the PU foam due to the aliphatic nature of its chains (ref).

Catalysts like dibutyltin dilaurate (DBTDL) are used to accelerate the reaction between the polyol and isocyanate. A small amount of water is introduced as a

Table 2.1 Types of lignin and their applications in polyol production

Type of lignin	Source	Characteristics	Applications in polyol production	References
Kraft lignin	Paper and pulp industry (Kraft process)	High sulfur content, stable, widely available	Suitable for producing rigid polyurethane foams	[53, 54]
Soda lignin	Soda pulping (non-wood plants like straw)	Sulfur-free, good water solubility, cleaner profile	Used in flexible foams and elastomers due to its sulfur-free nature	[55]
Organosolv lignin	Organic solvent extraction (e.g., ethanol)	High purity, sulfur-free, low molecular weight	Ideal for high-quality polyols used in both rigid and flexible foams	[56–58]
Lignosulfonates	Sulfite pulping process	High sulfur content, water-soluble, low-cost	Requires extensive modification, used in rigid foams or as cross-linkers	[59, 60]
Enzymatic hydrolysis lignin	Enzymatic hydrolysis of biomass	Less modified, preserves more of native lignin properties	Potentially useful for high-performance polyurethanes, good for sorbents due to retained properties of lignin	[61–64]

blowing agent to generate the foam's porous structure. When water reacts with isocyanate, it produces carbon dioxide gas, which creates the cellular structure characteristic of polyurethane foams. Suppose additional properties, such as magnetic responsiveness or solar-driven heating, are required. In that case, nanoparticles like Fe_3O_4 or carbon nanotubes (CNTs) are dispersed into the polyol mixture using ultrasonication before the isocyanate is added. This ensures uniform distribution of the nanoparticles, enhancing the foam's functional properties.

And lastly, To optimize lignin-based polyurethane foams for specific tasks like oil–water separation, post-production surface modifications are often applied. For applications requiring high oil absorption, the foam can be rendered superhydrophobic by treating it with octadecyltrimethoxysilane (OTMS). This treatment increases the water contact angle of the foam, making it highly water-repellent and ideal for selectively adsorbing oil from water. Additionally, when Fe_3O_4 magnetic nanoparticles are incorporated into the foam, the material becomes magnetically responsive, allowing for easy recovery and reuse by applying a magnetic field.

In contrast, some foams are modified to achieve superhydrophilic and underwater superoleophobic properties, making them highly effective at oil–water separation in aqueous environments. This is done by coating the foam with polydopamine (PDA) to increase surface roughness and enhance adhesion. The foam is then treated with phytic acid (PA), which makes the surface hydrophilic, allowing water to pass through while repelling oil. This dual modification ensures that the foam remains free of oil contamination, prolonging its service life in separation applications.

One of the advantages of polyurethane-based sorbents, particularly those made from bio-based materials like lignin, is that they present a truly sustainable alternative to traditional, petroleum-based PU sorbents. Not only do they help reduce our reliance on fossil fuels, but they also offer comparable performance, making them an eco-friendly choice for various applications.

2.2.2 Cellulose-Based Foams

Cellulose is Earth's most abundant natural polymer, derived primarily from plant cell walls. It comprises long chains of glucose units, forming a highly organized structure that gives plants strength and rigidity. Due to its chemical structure, cellulose-based sorbents have been studied recently. The ability of cellulose to be chemically modified has led it to be used for a wide array of applications. It can be selectively modified to create a hydrophobic/oleophilic or a hydrophilic/oleophobic material, creating products that can be used depending on a specific purpose.

There are several ways to modify cellulose. Grafting has been a widely used approach to modify cellulose to fit its intended properties; for example, grafting materials such as poly(hydroxamic acid) ligands to cellulose was found to have an excellent absorptive capacity of toxic metal ions [65], and Butyl methacrylate was grafted to cellulose to create an oil absorbent and was found to have 3–6 times as compared to that of a commercial polypropylene sorbent [66].

Cellulose is naturally hydrophilic due to its structure, composing of multiple hydroxyl groups. Because of this, reactions have been taken to reduce its OH groups. Jarrah et al. [67] reports grafting alkyl groups into the cellulose structure through a silylation reaction as the silylation reaction converts the OH groups to alkyl silyl ethers, which increases the hydrophobicity of the sorbent. There are other methods, such as acetylation, to react and convert the hydroxyl groups of the cellulose to acetyl groups ($–COCH_3$), which increases the hydrophobicity/oleophilicity of the fibers [67, 68]. Undoubtedly, sorbents derived from cellulose show potential for pollution control and waste management, especially for oil spills.

In recent years, cellulose-based sorbents for oil spills have garnered attention. Several sorbents have been synthesized for oil spill cleanup. One widely studied form is cellulose-based aerogels; these materials have good hydrophobicity and great oleophilic properties, are lightweight and compact, and generally have good sorption capacity [69]. However, their downside is their reduced mechanical properties and chemical instability, limiting their application [70].

In recent innovations, cellulose has been harnessed to create foams such as Water-stable cellulose fiber foams for packaging [71], insulation foams [72], and with the advent of sorbent foams, shows promise for oil spills. Sorbents in the form of foams provide a high surface area and porosity, enabling efficient absorption of different pollutants.

Foams derived from cellulose undergo different processes to be synthesized. Different forms of cellulose, such as dissolved cellulose and nanocellulose, are used to create the foams. First, these materials are suspended in a solvent, and the cellulose is dispersed, stabilizing the network structure with hydrogen bonds or chemical crosslinks. After this process, the foams need to be dried, which can either be freeze-dried, where the suspension is frozen and then the solvent is sublimed under a vacuum, leaving the foam, or dry-dried, which can be freeze-dried, where the suspension is frozen and then the solvent is sublimed under a vacuum, leaving the foam. The resulting structure is a porous, lightweight foam with a high surface area. In the case of cellulose-based foams, freeze drying is mostly preferred due to its ability to retain the 3D structure of the material, however, the downside is that it is time consuming. Recently, convective (or air drying) has emerged as to replace the traditional drying, offering a less complex way of producing these materials [73]. Figure 2.5 summarizes the general method of producing cellulose-based foam sorbents.

However, as mentioned previously, cellulose is inherently hydrophilic. So to create sorbent foams for oil spills, it must chemically modified, be it through grafting or

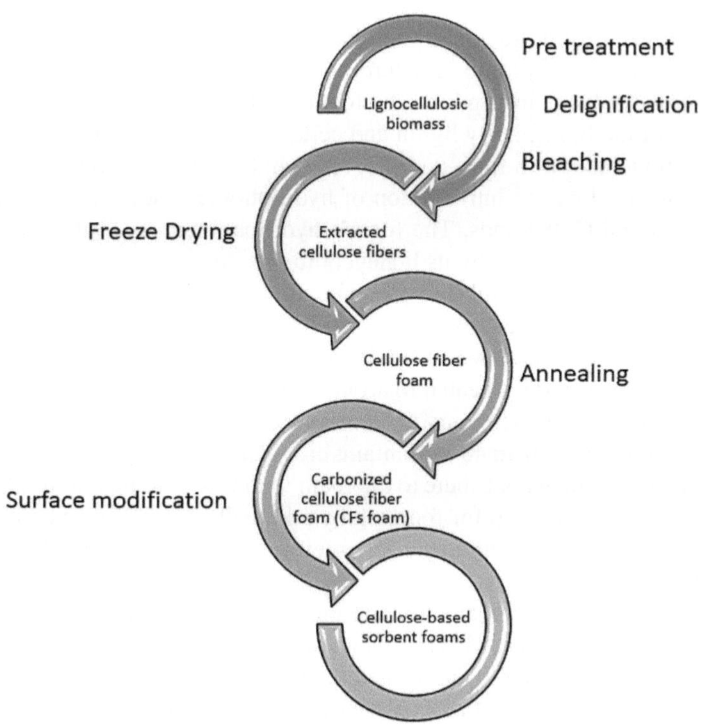

Fig. 2.5 Schematic of producing the cellulose based foam. Modified from Ma et al. [74]

other methods. For example, the study of Ma et al. [74], they produced a multi-functional cellulose-based foam through modifying biomass carbon fibers through silylation using polydimethylsiloxane (PDMS) that is immobilized with silver oxide. The PDMS is used to create a hydrophilic/lipophilic structure, as the siloxane groups reacts with the hydroxyl groups, undergoing a silylation reaction, and grafting the PDMS onto the cellulose. The sorbent foams created from this had good properties such as high hydrophobicity with a water contact angle of 147.8° and an excellent absorption of capacity of heavy crude oil of around 60.6 g/g.

This proves that cellulose has the potential for oil spill cleanups. Its lightweight, high surface area, and highly modifiable nature make it a good alternative to conventional sorbents, enabling the development of sustainable sorbents for oil spills.

2.2.3 Larch Saw Dust-Based Foams

As of the time this book was written, only 1 study explored larch sawdust-based polymer foam by Tan et al. [75], which has a different approach in making the foams compared to its lignocellulosic counterparts. The carbon foam produced from larch sawdust—specifically referred to as liquefied-larch-based polymer foam (LLB-PF) and its carbonized form, liquefied-larch-based carbon foam (LLB-CF)—exhibits a 3D interconnected honeycomb structure, with open-cell morphology that enables efficient oil and solvent absorption. The foam synthesis capitalizes on the chemical components of larch, primarily lignin and cellulose, which decompose and restructure during liquefaction to form a stable, cross-linked polymer matrix. This matrix is further enhanced by the introduction of hydrophobic functional groups, such as benzene rings and C=C bonds. The foam's hydrophobicity is further amplified by the high surface area created by its honeycomb structure.

The synthesis of honeycomb-like carbon foam from larch sawdust begins with the liquefaction of larch sawdust using phenol and formaldehyde in an alkaline solution, which promotes the breakdown of lignin and cellulose into reactive multiphenolic compounds. This mixture is heated to around 55–60 °C, then to 95–98 °C, allowing resinification to occur, which transforms the sawdust into a homogeneous, polymer-like solution. The resulting material contains cross-linked multiphenolic rings derived from the sawdust, which contribute to the foam's stability and reactivity. Once liquefied, the solution is prepared for foaming by adding Tween 80 as an emulsifier and n-pentane as a foaming agent, ensuring an even distribution of bubbles within the polymer matrix.

Following the preparation, the mixture undergoes a foaming and curing process. Sulfuric acid is introduced as a curing agent, initiating a rapid foaming reaction as the solution is poured into a mold and heated to 60 °C. During this stage, the n-pentane volatilizes, forming bubbles that coalesce into a stable, interconnected, open-cell structure. The result is a lightweight, honeycomb-like polymer foam (LLB-PF) with structural integrity and a highly porous surface ideal for absorption. This stage allows the polymer to form large interconnected cells for oil absorption while maintaining low density due to the minimal polymer structure.

To produce the final carbon foam, the polymer foam (LLB-PF) is carbonized in a nitrogen-purged furnace, heating it to 900 °C. The carbonization removes volatile organic components, stabilizing the remaining carbon structure and shrinking the cells slightly to increase durability. This carbonized form, known as LLB-CF, retains the honeycomb open-cell morphology and exhibits high hydrophobicity and oleophilicity due to the carbon-based structure with nonpolar functional groups. This combination of structure and chemistry results in a durable, hydrophobic foam with a substantial surface area that is ideal for selectively absorbing oils and organic solvents while repelling water.

Several tests were performed to assess the compatibility of the foam and determine whether this would be a viable sorbent material. Such tests performed were Scanning Electron Microscopy (see Fig. 2.2, something), Transmission Electron Microscope (TEM), Fourier Transform Infrared spectroscopy (FTIR), Thermogravimetric Analysis (TGA), and a contact angle system (DSA25). Tests such as recyclability and separation were performed to assess the material's performance compared to its petroleum counterparts (more of their properties in Chaps. 3 and 4). As it turns out, the produced materials have an absorptive capacity of up to 88–153 times for oils and organic solvents, and they also showed excellent absorption capacities even after five cycles.

With the advent of this innovation, further research could potentially expand and harness the properties of this honeycomb-like carbon foam, optimizing its structure and absorption capacity for a broader range of environmental applications, including more efficient cleanup of various organic pollutants and enhanced recyclability in industrial spill responses.

2.2.4 Alginate-Oil Gelator Foam Composite

The development of alginate-based foam composite sorbents has garnered significant attention for their potential application for oil spill remediation. Alginate is a biomaterial typically extracted from brown algae (Phaeophyceae). To create ionic bonding with uronic acid residues, it can interact with divalent or trivalent metal ions to construct a three-dimensional network using an "egg-box" model [76]. It is treated with aqueous alkali solutions, filtered, and added with sodium or calcium chloride to precipitate the alginate. It is also a biopolymer with properties, such as its ability to form hydrogels, biocompatibility, and capacity for ion exchange, that make it beneficial for environmental remediation, specifically foam sorbents, biomedical fields, and others specified in Fig. 2.6 [77]. The mechanical properties of alginate, usually formed as gels, are influenced by its cross-linking density and chemical composition, which can be tailored to optimize swelling behavior and mechanical strength [78]. This tunability is crucial for developing foam sorbents that capture and retain contaminants.

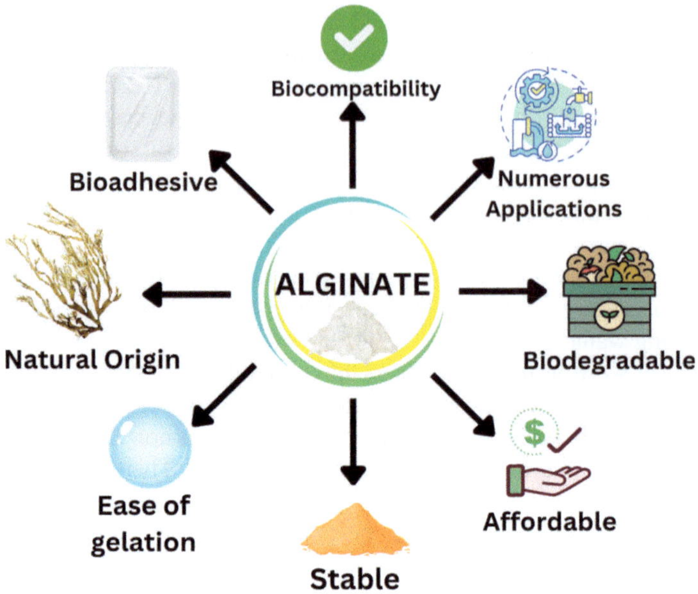

Fig. 2.6 Adapted from schematic presentation of alginate properties, Adamiak and Sionkowska [79]

Alginates primarily consist of hydrophilic molecules due to the presence of carboxyl and hydroxyl groups. Imparting hydrophobicity to the alginate involves chemical modifications done in the hydroxyl groups, which involve reductive amination, sulfation, and oxidation. At the same time, esterification and amidation reactions take place in the carboxyl groups [80]. Other than modifying the hydroxyl and carbonyl groups through chemical reactions, techniques such as chemical vapor deposition, salinization, freeze-drying, crosslinking, alkyl grafting, and incorporation of essential oils can be incorporated [81]. The properties of alginates are crucial for creating alginate-based foams for oil spill remediation where the development of alginate-oil composite foam has emerged as a significant innovation. Figure 2.7 illustrates the use of alginate's property to form hydrogels through ionic gelation, particularly with divalent cations such as calcium ions, to synthesize alginate-based foams. To increase the surface area of the composite foam available for oil absorption, CO_2 is incorporated into the porous structure during the foaming process to facilitate foaming and gelation simultaneously. The resulting foam exhibits a three-dimensional network that effectively traps oil [82].

One emerging technology for addressing oil spills is the development of alginate-oil gelator composite foam. Synthesizing composite foams involves combining a hydroxyl aluminum stearate oil gelator with an alginate foam matrix to enhance the foam's oil absorption capacity significantly [83]. The synthesis of the alginate foam oil absorbent involves freeze-drying and hydrophobic modification. The sodium alginate solutions are prepared first by dissolving the alginate powder in distilled

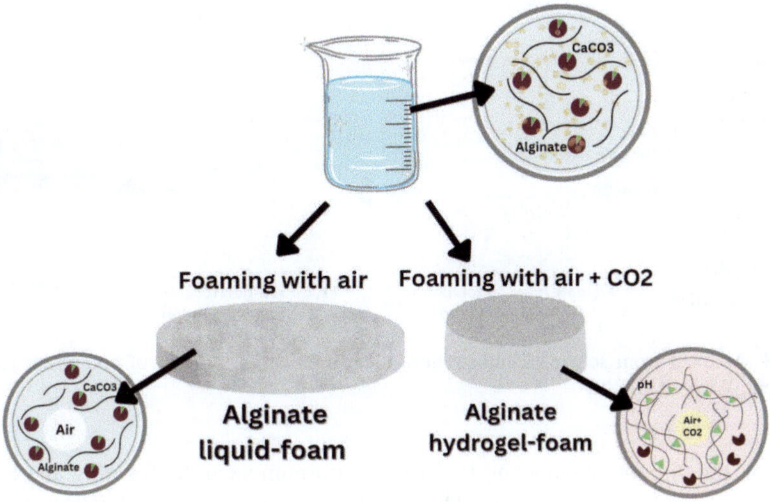

Fig. 2.7 Adapted from graphical abstract, Djemaa et al. [83]

water or t-butanol. Introducing t-butanol improves the porosity and enhances the hydrophobicity of the resulting foam, further impressing its enhanced properties [84]. The solutions are left at room temperature overnight to ensure complete dissolution of the alginate. The solutions are then subjected to freeze-drying to remove the solvent from the solution and create a porous foam structure. The freeze-drying process is critical as it preserves the integrity of the foam while enhancing the surface area [85]. After freeze-drying, it undergoes further hydrophobic modification where fluorinated alkyl silanes (FAS) enhance the alginate foams' hydrophobic properties. This process involves placing the foam in a vacuum desiccator with FAS, allowing for a chemical reaction that introduces hydrophobic groups onto the alginate surface [86].

Oil gelators in foam sorbents can improve the hydrophobicity and oleophilicity of the material. For instance, aluminum hydroxyl stearate is commonly used as an oil gelator. Recent research had highlighted aluminum hydroxyl stearate as an effective gelator due to its gelation properties that allow it to effectively thicken hydrocarbons, forming a rigid shell that prevents oil efflux and facilitates cleanup efforts [87]. One function of oil gelators in polyurethane foams is that they facilitate the formation of a three-dimensional network within the foam structure, which contributes not only to the increase of surface area for enhanced oil absorption but also to the enhancement of the foam's mechanical stability [88]. Aluminum hydroxyl stearate is prepared by dissolving stearic acid into distilled water and then adding sodium hydroxide with vigorous stirring to produce a sodium salt solution. This reaction is continued for about an hour, and after washing and drying, white powder or granular products will be obtained [89]. The alginate foam oil absorbent and oil gelator solutions are combined through vigorous stirring, and aluminum sulfate is added to the solution to

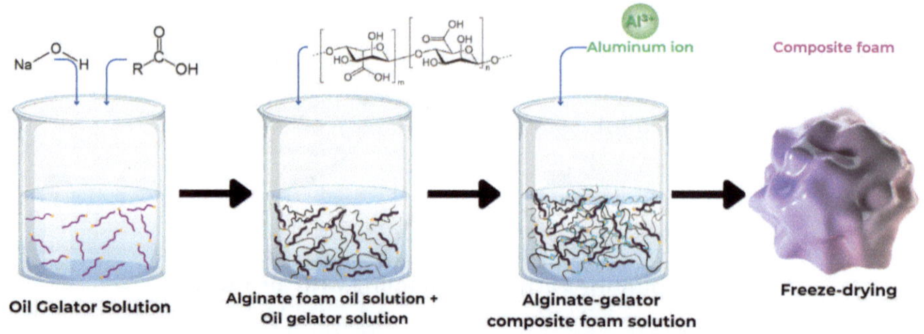

Fig. 2.8 Adapted from schematic illustration of the fabrication process of composite foam oil absorbent, Wang et al. [89]

crosslink the alginate chains and hydroxyl aluminum stearate, forming an alginate-gelator composite foam solution. This solution undergoes lyophilization, a freeze-drying process that removes moisture and solidifies the foam structure, forming composite foams with enhanced oil absorbency and stability described in Fig. 2.8 [89].

The properties of the composite foam are influenced mainly by the foam structure and oil gelator content. At 43% oil gelator content, including t-butanol (TBA) in the solvent significantly impacts the porous structure and surface hydrophobicity. As TBA content increases, from 12.5 to 37.5%, the foam develops a multiscale hierarchical structure, enhancing hydrophobicity with a contact angle increase from 126° to 154°. This improvement is attributed to the rougher surface and higher specific surface area. Optimal hydrophobicity and stability were achieved with 37.5% TBA and 43% oil gelator. The foam demonstrated selective oil absorption for various oils (e.g., hexane, chloroform) and organic solvents without swelling, absorbing up to 63 times its weight. Additionally, the oil gelator aids in forming stable oil gels within the foam, improving oil retention under pressure, with retention rates close to 90% compared to the alginate foam, which is 25% [89]. The foam's reusability was validated through oil absorption-extraction cycles, maintaining effective performance even after multiple uses.

2.2.5 *Polylactic Acid-Based Foam*

Polylactic acid (PLA), another raw material used for foam sorbents, is an aliphatic polyester composed of lactic acid or 2-hydroxypropanoic acid derived from renewable resources like corn starch and sugar cane. Its good biocompatibility, mechanical strength, and biodegradability make it a sustainable choice for various applications, such as the food packaging industry, automotive, and biomedical applications demonstrated in Fig. 2.9 [90]. Today, PLA is not just a raw material for bio-based polymers,

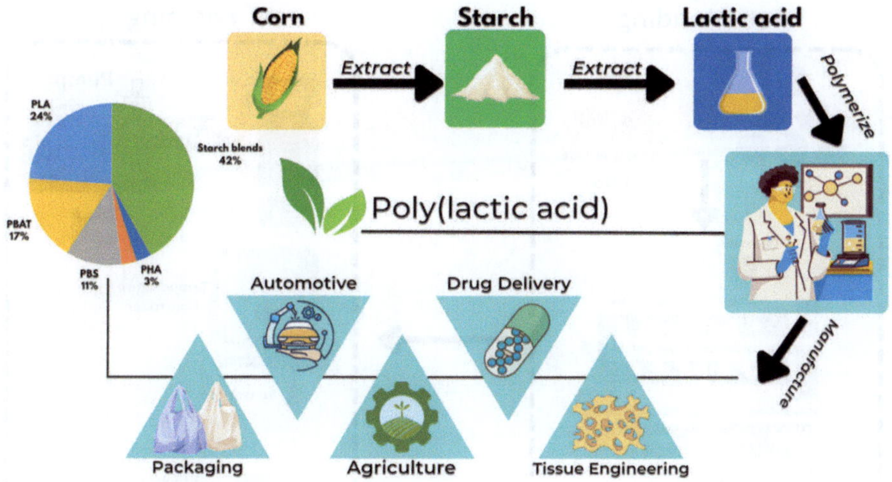

Fig. 2.9 Adapted from graphical abstract, Balla et al. [90]

but also a key player in the creation of foam sorbents for oil spill cleanup. The adaptability of PLA-based foams in oil spill remediation, with their potential for high porosity and surface area, instills versatility and effectiveness in oil absorption.

Research has shown that PLA foams can be fabricated using various methods, including supercritical CO_2 foaming, which enhances the foam's expansion and porosity. The synthesis of PLA foams using the supercritical fluid foaming process faces several challenges, particularly in achieving an open-cell foam structure with a high expansion ratio. Ren et al. found that the crystallization and phase separation in PLA/polyethylene glycol (PEG) blends restricted cell growth, leading to closed-cell structures rather than the desired open-cell foams [91]. The stiffness of the PLA matrix and its crystallinity can hinder the expansion of cells during foaming, resulting in lower expansion ratios and less effective oil absorption capabilities. Li et al. reported super high-expansion PLA foams that exhibit outstanding selective oil-absorption performance. These involve a PLA/poly(butylenes succinate) (PBS) foam blend, where PLA and PBS pellets are dried, and then melt blended using a twin screw extruder with a screw speed to ensure uniform mixing of the two polymers. The blend undergoes an extrusion process where the polymers are melted and mixed together under high temperature and shear forces to create a homogenous blend. The PLA/PBS blend uses a supercritical CO_2 foaming process, a solvent-free process, to effectively create open cell structures shown in Fig. 2.10. This technique allows for the generation of high porosity and large surface areas, which are critical for effective oil adsorption [92]. The PLA/PBS foam demonstrates a highly interconnected open-cell structure with 98.2% open-cell content and a 43.6 expansion ratio, resulting in high porosity (97.7%). This structure allows the foam to selectively adsorb oils and organic solvents, with an adsorption capacity ranging from 7.9 to 21.9 g/g while maintaining stable performance over 20 adsorption–desorption cycles for CCl_4 [92].

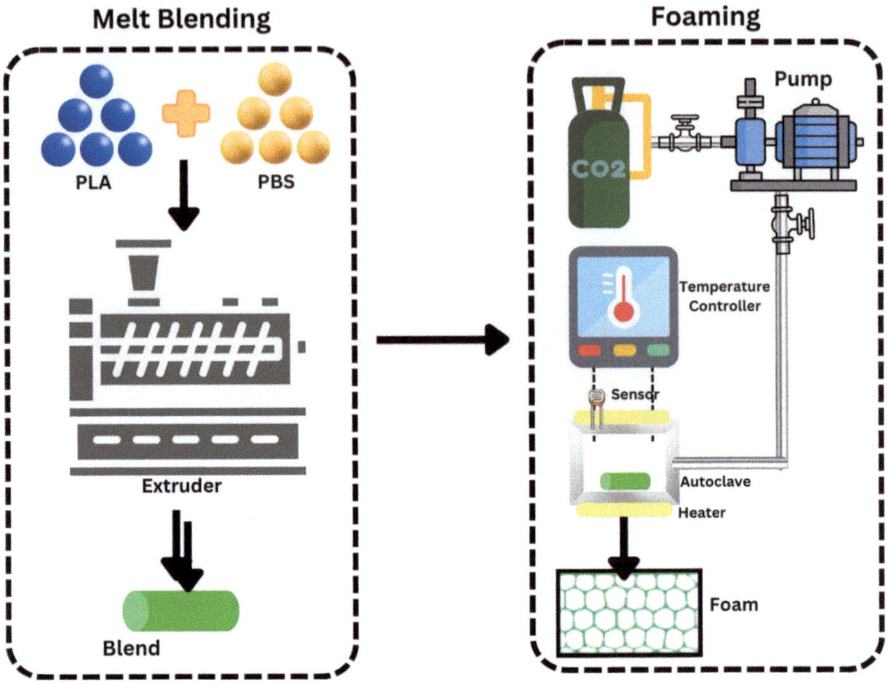

Fig. 2.10 Adapted from schematic diagram of the preparation process of the PLA/PBS foams, Li et al. [92]

Another biodegradable foam sorbent incorporating supercritical CO_2 has been discovered. This presents a significant advancement in the development of efficient oil–water separation. Yang et al. demonstrate the fabrication of PLA bio-based sorbent facilitated by poly(3-hydroxybutyrate-co-3-hydroxyvalerate) (PHBV) to have a bio-based, biodegradable, and superior oil–water separation. PHBV is a copolymer that can form interconnected open-cell structures essential for effective oil absorption. The open pores in PHBV foams facilitate better oil uptake through capillary action, making them suitable for applications in oil spill remediation [93]. The synthesis of PLA/PHBV composite foams were prepared by melt mixing different weight ratios of PLA and PHBV at 180 °C using an internal mixer, followed by compression molding at 190 °C under high pressure. The foaming process involved a two-step method where the samples were initially exposed to CO_2 at 6.1 MPa in an autoclave at 30 °C, followed by heating to 170 or 180 °C to partially or fully melt the crystalline regions. The foaming was completed at 120 °C (with data from 110 and 130 °C provided separately), and the samples were rapidly depressurized to form the foams, resulting in materials with open-cell structures and high volume expansion ratios [94].

The addition of PHBV enhances the hydrophobicity, with a contact angle reaching 134.5°, attributed to the open-cell structure and alkyl groups of PHBV. The foam's oil adsorption capacity increases with PHBV content, peaking at 5% PHBV. The PLA/PHBV5 foam effectively absorbs floating and submerged oils, making it suitable for oil–water separation with maximum adsorption capacity of 51.8 g/g for CCl_4. The adsorption capacity remains stable through multiple cycles, and the foams exhibit stronger adsorption for low-viscosity oils like CCl_4 compared to higher-viscosity liquids [94]. This strong oil adsorption behavior is maintained over multiple adsorption–desorption cycles due to a pseudo-second-order adsorption kinetic model, where fluidity and oil viscosity play a role in the rate of adsorption. The improved hydrophobicity is due to PHBV's alkyl groups and the open-cell foam structure, which increases surface roughness, reducing water contact and improving oil affinity. PHBV's hydrophobicity and crystallinity support this change, while the foam's porosity (open-cell content) allows for better oil absorption, particularly in PLA/PHBV5 blends with an optimal PHBV content [95].

The incorporation of natural fibers enhances the mechanical properties of composite foams, providing structural integrity crucial for oil absorption. Kaisone et al. highlighted that incorporating modified coconut fiber into PLA composite foams increases compressive and tensile strength [96]. Among natural fibers, kapok and hemp fibers have shown higher oil absorption capacity. Kapok fibers have been reported to exhibit excellent water repellency and high oil affinity due to their cellulose-based composition and hollow structure. They have thin-walled, hollow lumens coated with natural wax, making them highly effective as oil sorbents [97]. Similar to kapok fibers, Calotropis gigantea fiber (CGF) also has a natural wax coating and hollow structure, with the addition of a fibrous nature that facilitates better oil uptake through capillary action and adsorption mechanisms [98]. In the study of Xiao et al., CGF and PLA are used as the starting materials undergoing a tandem freezing-solvent replacement method to construct a foam-like oil sorbent. PLA foam (PF) and PLA-CGF foam (PCF) were both prepared using a freezing and solvent replacement method. For PF foams, varying amounts of PLA were dissolved in 1,4-dioxane, frozen at 18 °C, and then immersed in distilled water to replace the solvent before drying at 60 °C. Similarly, PCF foams incorporated 0.1 g of pre-cut CGF (< 3 mm) into the PLA solution before undergoing the same freezing and solvent replacement steps. The resulting foams were labeled according to their PLA content. The best-performing foams from each set were selected for further oil absorption experiments [99]. The PCF exhibited a stable porous structure and excellent mechanical strength, withstanding 185 kPa pressure. CGF's high cellulose content provides mechanical reinforcement and the fibers act as load carriers to PLA composite foams, effectively distributing stress throughout the composite during mechanical loading, enhancing the foam's overall structural integrity [100]. It also demonstrated significant oil-absorption capacities of 22.8–48.3 g/g and remained effective in high-temperature water environments due to its hydrophobic-oleophilic properties. PCF maintained reusability, with absorption percentages of 81.3% for soybean oil and 88.1% for kerosene after 10 cycles [99].

Oil-sorbent composite foams can also be improved in terms of performance by using nanoparticles. Montes et al. reported the addition of amorphous lignin nanoparticles (LNPs) to PLA films reduces crystallinity and improves flexibility, making

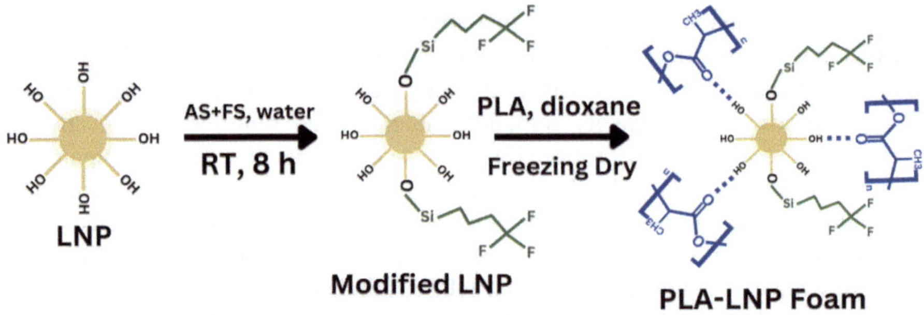

Fig. 2.11 Adapted from the preparation process of PLA composite foams, Ding et al. [102]

these materials suitable for applications involving durable materials. Incorporating LNPs could also distribute stress more evenly throughout the foam, increasing the tensile strength and impact resistance of the obtained foam [101]. Ding et al. discuss impregnating LNPs into biodegradable polymers as an environmentally friendly method for effective oil/organic solvent absorption. A bio-based PLA foam with enhanced hydrophobicity and a uniformly porous structure is achieved through a green, simple modification using LNPs combined with freeze-drying shown in Fig. 2.11. The synthesis of PLA/LNP composite foam involves dissolving PLA in 1,4-dioxane with varying loadings of modified LNPs (m-LNP) to produce a homogenous dispersion. The mixture was then frozen at − 20 °C overnight to allow the foam structure to form easily. Finally, PLA hybrid foams in different concentrations of LNPs were obtained through a freeze-drying process [102]. When 6 wt% of m-LNP were added to PLA foams, porosity increased slightly from 93.0 to 94.0%. Incorporating m-LNP reduced foam shrinkage and collapse during freeze-drying, attributed to the rigidity of the aromatic skeleton in m-LNP. This inclusion also resulted in a high increase in the hydrophobicity of the foam as the water contact angle increased from 124.8° in neat PLA to 141° for the PLA-6L sample [102]. Freeze-dried fluorinated m-LNP gave a highly hydrophobic surface and reduced the interaction with water with improved absorption with oil [103].

2.2.6 *Natural Rubber Based Foam*

Natural rubber (NR), a renewable polymer of cis-1,4-polyisoprene derived from the latex of rubber trees, is a versatile material known for its excellent mechanical properties, elasticity, and hydrophobic properties. Its applications span various industries, including automotive, construction, and medical. The unique characteristics of NR, such as elasticity and resilience, make it an ideal candidate for numerous applications, particularly in oil spill remediation [104, 105]. Studies by Chin et al. and Mustapa et al. pointed to the applicability of natural rubber foam (NRF) as an alternative to commercial sorbents; however, the use of NRF resulted in the oil absorption efficiency declining with further cross-linking agents, which decreased its general sorption capacity [106, 107]. The natural rubber industry reveals that reinforcing fillers are vital to obtaining desirable properties for numerous commercial applications. Traditional reinforcing fillers in the rubber industry include nonrenewable inorganic materials like carbon black, silica, calcium carbonate, and clay. Recent advancements in nanotechnology have led to the utilization of nano-sized renewable fillers to reinforce NR composites [108]. Study of Songsaeng et al. reported the use of reduced graphene oxide (rGO) to increase that natural rubber composites' surface area and oil absorption capacity, enhancing their effectiveness as sorbents. The hydrophobicity of rGO in the NR matrix increases interaction at the filler-matrix interface, affecting the mechanical, thermal, and oil sorption properties. rGO contributes to the large surface area toward better dispersion within the rubber due to strong interfacial bonding, which provides a more uniform structure and improves barrier properties [109].

The production of rGO forms through the oxidation of graphite waste in sulfuric acid with potassium permanganate, followed by deionized water addition and ultrasonic treatment. Sodium hydroxide adjusts the pH, while L-ascorbic acid completes the reduction. The final product, a black precipitate, undergoes filtration and washing before drying to yield rGO. To prepare NR/rGO composite foam sorbents, natural rubber latex is mixed with varying rGO concentrations and cured, molded, and washed. These composite foams then undergo testing to assess their effectiveness in oil absorption applications [104]. NR/rGO composite foams were developed as oil-sorbent materials, with their performance influenced by factors like foam morphology, rGO content, soil type, and properties. The foam's high porosity, especially at 0.5 phr rGO (NRG-0.5), optimizes oil absorption, with gasoline absorption at 21.50 g/g and crude AXL oil at 17.04 g/g. Gasoline's lower viscosity leads to more excellent selectivity due to faster diffusion. Increased temperatures up to 45 °C and external forces like waves further improve absorption. Adsorption follows the pseudo-second-order model, while the Langmuir isotherm suggests a monolayer adsorption mechanism. NRG-0.5 exhibits significant elasticity and durability, with which the material can be reused for at least 30 cycles without sustaining any damage, indicating that it might be a very promising solution for ocean oil spill cleanups [104].

Recently, silica-lignin has also emerged as an effective reinforcement filler for foam absorbent. Silica produced from rice husks has a significantly developed surface area and porosity, making it an efficient filler in rubber composites. Natural polymer lignin, found in plant cell walls, further enhances the composite's mechanical strength and thermal stability. The combination of silica and lignin also strengthens the rubber matrix's physical properties and helps meet sustainability requirements by utilizing agricultural waste. Several steps are involved in extracting silica and silica–lignin hybrid materials to form the absorbent form described at Fig. 2.12. Rice husk ash (RHA) is obtained by burning rice husks at a temperature of 700 °C for 6 h and is used for silica extraction. On the other hand, rice husk (RH) is the precursor for silica–lignin hybrids. Combining RHA or RH with 1 M HCl facilitates the removal of metallic impurities, which are then heated, stirred, and filtered. After neutralization,

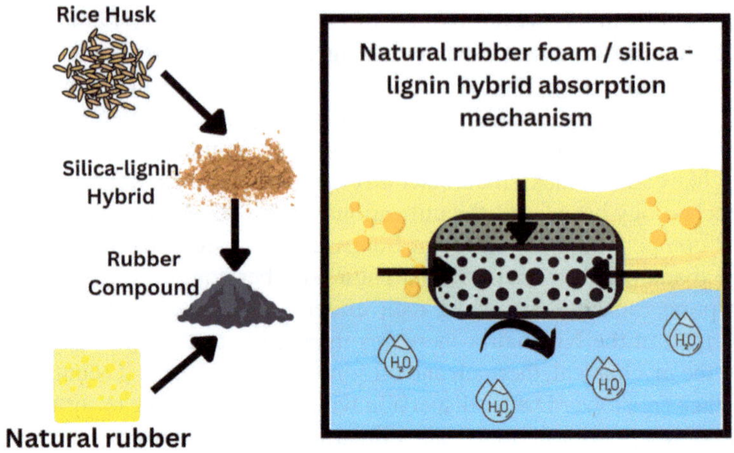

Fig. 2.12 Adapted from graphical abstract, Mardiyati et al. [110]

combining the HCl-treated material with 2 M NaOH leads to further extraction. The final silica precipitation occurs with the addition of 2 M H_2SO_4 until reaching a specific pH, followed by filtration and drying at 50 °C to yield the end product. The incorporation of the silica-lignin filler occurs during the compounding of natural rubber. To prepare the rubber compound, an XK-160 open mill operates at a friction ratio of 1:1.35 and a roll speed of 13.39 rpm [110]. These NRF samples showed good hydrophobicity with a contact angle of 133° for water. The reusability was found to be more stable than both the unfilled NRF and silica-filled NRF. Additionally, the maximum amount of oil absorbed by the silica-lignin hybrid-filled NRF was observed to be up to 1.36 g/g. The performance of the silica-lignin hybrid-filled NRF appears to have good potential as a sustainable oil-absorbent material with economic and environmental benefits [110].

2.2.7 Tannin-Based Polyurethane Foam

Tannins, complex large biomolecules of natural polyphenolic compounds derived from various plant species, have emerged as a sustainable primary raw material for synthesizing sorbents. They are categorized as condensed and hydrolyzable tannins, with condensed tannins being the dominant ones in the market. Condensed tannins are flavolans or polymers of flavan-3-ols (catechins) and flavan 3:4-diols (leucoanthocyanidins) that can react with aldehydes to form polymeric materials that make them valuable in fields such as wood treatment, leather processing, packaging, 3D printing, etc. [111].

Due to their rich hydroxyl groups and rigid structure, tannins can be utilized as key raw materials to create polyurethane foams. Polyurethane foam (PUF) has high porosity and large surface area contributing to a high absorption capacity [112]. However, water absorption presents challenges for PUFs in oil–water separation applications. Hydrophobic modifications, incorporating materials like polydimethylsiloxane [113], carbon nanotubes [114], and rGO [115], enhance the properties of PUFs, but the preparation process is complex and time-consuming. On the other hand, tannin-based polyurethane foams can be synthesized relatively straightforwardly by mixing tannins with polyols, foaming agents, or other crosslinking agents. This was demonstrated in the study of Li et al., which used glyoxal as a crosslinking agent to improve the properties of the tannin-based PUF [116]. This ease of synthesis allows for the efficient production of foams tailored to specific applications.

In Zhang et al.'s study (2022), tannin extracts are mixed with propylene oxide to create a tannin-based polyol through an esterification reaction. These polyols are then used to produce modified tannin-based polyurethane foams (MTPFs) for use as absorbent materials shown in Fig. 2.13. The tannin structure plays a crucial role in the final foam properties, providing both rigidity and functionality for efficient oil absorption. The tannin is extracted from waste bark, which is cheap and readily available, through liquefaction. The biomass is mixed with a solvent like DMF (dimethylformamide) and a catalyst like KOH (potassium hydroxide) in a reactor. The system is purged with an inert gas, such as nitrogen, to maintain a controlled atmosphere, and the mixture is stirred and heated to a specified temperature range. A reagent such as propylene oxide is gradually introduced into the reactor over a set period. After the reaction, the mixture is cooled, and volatile components are removed. The resulting crude product is neutralized, treated with adsorbents like activated clay, filtered, and concentrated under reduced pressure to yield a liquid product with desired viscosity and yield [117]. The foam preparation involves mixing the polyol components in 200 mL containers at 2000 rpm for 5 min. Then, add polymeric MDI and stir the mixture for 10 s to initiate foam formation. The resulting foams contained varying levels of the tannin-based polyol, replacing 0–100 wt% of SA-380 (polyether polyol), and were labeled accordingly. Leave the samples to age for 48 h, and measure the key foaming parameters such as cream, rise, and dry times [117].

MTPFs improve performance in several key areas compared to unmodified foam (RPUF). Water absorption decreased from 7.7% in RPUF to 6.4–7.2% in MTPFs, reflecting improved hydrophobicity. This was further supported by an increased contact angle, rising from 98.3° in RPUF to 118.9–128.5° in MTPFs. The enhanced hydrophobicity can be attributed to the hydrophobic groups introduced into the polyurethane structure via tannin extract after liquefaction. Additionally, the apparent density of MTPFs increased from 32.5 to 35.5 kg/m^3, which can be linked to branched segments in the polyurethane, causing higher polymer macromolecule accumulation [117]. When interacting with isocyanates and polyols, tannins can form hydrogen bonds that enhance the cross-linking densities within the polymer matrix. This leads to a more stable and hydrophobic structure, as the tightly bound network reduces the mobility of water molecules within the foam [119]. Hydrophobic tannin moieties can also create a barrier against water penetration, enhancing the foam's hydrophobic properties [120]. Regarding performance, MTPF100 exhibited significantly enhanced adsorption capacities, with an increase of up to 115.1%

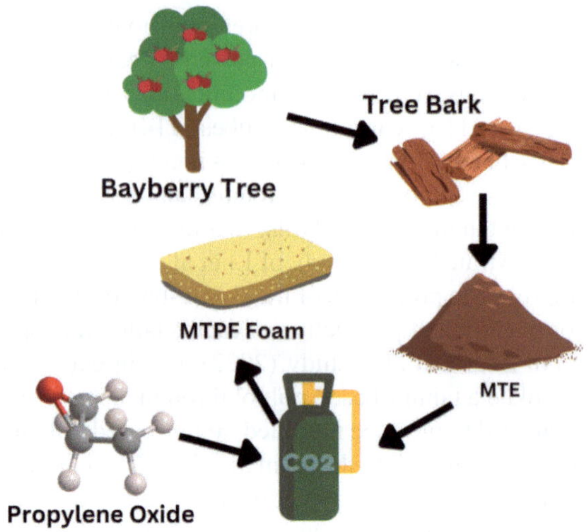

Fig. 2.13 Adapted from "liquefaction process of MTE and preparation of MTPFs foam," Li et al. [118]

for dichloromethane and notable improvements for soybean oil, xylene, n-hexane, and cyclohexane. Moreover, MTPF100 demonstrated better mechanical properties, improved thermal stability with an increased initial degradation temperature (209.5 °C compared to 121.4 °C for RPUF), and could be reused for six cycles without performance degradation. The presence of aromatic rings in tannins contributes to the foams' thermal stability and mechanical integrity, making them suitable for harsh environmental conditions [121, 122]. These advancements suggest that MTPFs are a promising eco-friendly and sustainable material for oil spill clean-ups.

References

1. Pfister, D.P., Xia, Y., Larock, R.C.: Recent advances in vegetable oil-based polyurethanes. Chemsuschem **4**(6), 703–717 (2011). https://doi.org/10.1002/cssc.201000378
2. Munir, M.A., et al.: Design and synthesis of conducting polymer bio-based polyurethane produced from palm kernel oil. Int. J. Polym. Sci. **2022**, Article ID 6815187, 1–13 (2022). https://doi.org/10.1155/2022/6815187
3. Miao, S., et al.: Soybean oil-based polyurethane networks: shape-memory effects and surface morphologies. J. Am. Oil. Chem. Soc. **90**(9), 1415–1421 (2013). https://doi.org/10.1007/s11746-013-2273-5
4. De Souza, V.H.R., Silva, S.A., Ramos, L.P., Zawadzki, S.F.: Synthesis and characterization of polyols derived from corn oil by epoxidation and ozonolysis. J. Am. Oil. Chem. Soc. **89**(9), 1723–1731 (2012). https://doi.org/10.1007/s11746-012-2063-5
5. Alaa, M.A., Yusoh, K., Hasany, S.: Comparative study of physico-chemical properties of pure polyurethane and polyurethane based on castor oil. Adv. Mater. Res. **983**, 39–43 (2014). https://doi.org/10.4028/www.scientific.net/amr.983.39
6. Liu, K., et al.: Castor oil-based waterborne polyurethanes with tunable properties and excellent biocompatibility. Eur. J. Lipid Sci. Technol. **118**(10), 1512–1520 (2016). https://doi.org/10.1002/ejlt.201500595

7. Kalita, H., Karak, N.: Biobased hyperbranched shape-memory polyurethanes: effect of different vegetable oils. J. Appl. Polym. Sci. **131**(1), Article ID 39579 (2013). https://doi.org/10.1002/app.39579
8. Chen, L., et al.: Synthesis of self-healing soybean oil-based waterborne polyurethane based on Diels–Alder reaction. J. Appl. Polym. Sci. **139**(30), Article ID e52694 (2022). https://doi.org/10.1002/app.52694
9. Ciastowicz, Ż, Pamuła, R., Białowiec, A.: Utilization of plant oils for sustainable polyurethane adhesives: a review. Materials **17**(8), Article ID 1738 (2024). https://doi.org/10.3390/ma17081738
10. Rojek, P., Prociak, A.: Effect of different rapeseed-oil-based polyols on mechanical properties of flexible polyurethane foams. J. Appl. Polym. Sci. **125**(4), 2936–2945 (2012). https://doi.org/10.1002/app.36500
11. Karimi, M.B., Khanbabaei, G., Sadeghi, G.M.M.: Vegetable oil-based polyurethane membrane for gas separation. J. Membr. Sci. **527**, 198–206 (2017). https://doi.org/10.1016/j.memsci.2016.12.008
12. Liu, W., et al.: Carbon nanofibers reinforced soy polyol-based polyurethane nanocomposites. J. Therm. Anal. Calorim. **123**(3), 2459–2468 (2015). https://doi.org/10.1007/s10973-015-4690-1
13. Şolea, L.C., Deleanu, L.: Flammability tests on hot surface for castor oil. Mech. Test. Diagn. **10**(4), 30–34 (2021). https://doi.org/10.35219/mtd.2020.4.04
14. Zhang, C., et al.: Soy-castor oil based polyols prepared using a solvent-free and catalyst-free method and polyurethanes therefrom. Green Chem. **15**(6), 1477–1484 (2013). https://doi.org/10.1039/c3gc40531a
15. Cai, W., et al.: Synthesis of bio-castor oil polyurethane flexible foams and the influence of biotic component on their performance. J. Polym. Res. **22**(8), Article ID 150 (2015). https://doi.org/10.1007/s10965-015-0782-7
16. Lee, J.H., Kim, S.H., Oh, K.W.: Bio-based polyurethane foams with castor oil based multifunctional polyols for improved compressive properties. Polymers **13**(4), Article ID 576 (2021). https://doi.org/10.3390/polym13040576
17. Shaik, A.H., et al.: Foam stability and thermo-mechanical properties of micro/nano filler loaded castor oil based flexible polyurethane foam. Mater. Res. Express. **11**(1), Article ID 015302 (2024). https://doi.org/10.1088/2053-1591/ad19b1
18. Chandan, M.R., et al.: Foam stability and polymer phase morphology of flexible polyurethane foams synthesized from castor oil. J. Appl. Polym. Sci. **131**(17), Article ID 40668 (2014). https://doi.org/10.1002/app.40668
19. Kaur, R., Kumar, M.: Function of silicon oil in the castor oil based rigid polyurethane foams. J. Polym. Eng. **33**(9), 875–880 (2013). https://doi.org/10.1515/polyeng-2013-0083
20. Chen, Y., Tai, W.: Castor oil-based polyurethane resin for low-density composites with bamboo charcoal. Polymers **10**(10), Article ID 1100 (2018). https://doi.org/10.3390/polym10101100
21. Çaylı, G., Küsefoğlu, S.: Biobased polyisocyanates from plant oil triglycerides: synthesis, polymerization, and characterization. J. Appl. Polym. Sci. **109**(5), 2948–2955 (2008). https://doi.org/10.1002/app.28401
22. Milanese, A., Cioffi, M., Shigue, H., Voorwald, C.: Cure kinetic of castor oil-based polyurethane. J. Appl. Polym. Sci. **122**(5), 3168–3171 (2011). https://doi.org/10.1002/app.34333
23. Wang, J., Zhang, C., Deng, Y., Zhang, P.: A review of research on the effect of temperature on the properties of polyurethane foams. Polymers **14**(21), Article ID 4586 (2022). https://doi.org/10.3390/polym14214586
24. Son, Y., Kim, J., Choi, S., Kim, J., Her, N., Lee, J.: Synthesis of polyurethane foam considering mixture blowing agents for application to cryogenic environments. Macromol. Mater. Eng. **304**(11), Article ID 1900294 (2019). https://doi.org/10.1002/mame.201900294
25. Alves, F.C., et al.: Efficiency of castor oil–based polyurethane foams for oil sorption S10 and S500: influence of porous size and statistical analysis. Polym. Polym. Compos. **29**(9_suppl), S1063–S1074 (2021). https://doi.org/10.1177/09673911211040360

26. Perera, H., Mortazavian, H., Blum, F.: Surface properties of silane-treated diatomaceous earth coatings: effect of alkyl chain length. Langmuir **33**(11), 2799–2809 (2017). https://doi.org/10.1021/acs.langmuir.7b00015
27. Wang, C., et al.: Castor oil-based polyurethane/silica nanocomposites: morphology, thermal and mechanical properties. Polym. Compos. **39**(S3), E2240–E2247 (2018). https://doi.org/10.1002/pc.24798
28. Kim, H.J., Jin, X., Choi, J.W.: Investigation of bio-based rigid polyurethane foams synthesized with lignin and castor oil. Sci. Rep. **14**(1), Article ID 10552 (2024). https://doi.org/10.1038/s41598-024-64318-8
29. Sharma, C., et al.: A pre-polyaddition mediation of castor oil for polyurethane formation. J. Appl. Polym. Sci. **133**(38), Article ID 43964 (2016). https://doi.org/10.1002/app.43964
30. Palanisamy, A.: Water-blown polyurethane–clay nanocomposite foams from biopolyol—effect of nanoclay on the properties. Polym. Compos. **34**(8), 1306–1312 (2013). https://doi.org/10.1002/pc.22543
31. Perera, H.J., Goyal, A., Alhassan, S.M., Banu, H.: Biobased castor oil-based polyurethane foams grafted with octadecylsilane-modified diatomite for use as eco-friendly and low-cost sorbents for crude oil clean-up applications. Polymers **14**(23), Article ID 5310 (2022). https://doi.org/10.3390/polym14235310
32. Suresh, K.: Rigid polyurethane foams from cardanol: synthesis, structural characterization, and evaluation of polyol and foam properties. ACS Sustain. Chem. Eng. **1**(2), 232–242 (2013). https://doi.org/10.1021/sc300079z
33. Kirpļuks, M., et al.: High functionality bio-polyols from tall oil and rigid polyurethane foams formulated solely using bio-polyols. Materials **13**(8), Article ID 1985 (2020). https://doi.org/10.3390/ma13081985
34. Amorim, F.V., et al.: Development of hydrophobic polyurethane/castor oil biocomposites with agroindustrial residues for sorption of oils and organic solvents. J. Colloid Interface Sci. **581**, 442–454 (2021). https://doi.org/10.1016/j.jcis.2020.07.091
35. Maia, L.S., Zanini, N.C., Claro, A.M., Amaral, N.C., Barud, H.S.: Eco-friendly foams of castor oil-based polyurethane with artemisia residue fillers for discarded vegetable oil sorption. J. Appl. Polym. Sci. **138**(43), Article ID e51259 (2021). https://doi.org/10.1002/app.51259
36. Costa, I.L., Monticeli, F.M., Mulinari, D.R.: Polyurethane foam reinforced with fibers pineapple crown biocomposites for sorption of vegetable oil. Fibers Polym. **21**(8), 1832–1840 (2020). https://doi.org/10.1007/s12221-020-9979-4
37. Qiu, Q., et al.: Effect of fiber surface treatment on the structure and properties of rigid bagasse fibers/polyurethane composite foams. Polym. Compos. **42**(6), 2766–2773 (2021). https://doi.org/10.1002/pc.26011
38. Charoensopa, K., Thangunpai, K., Kong, P., Enomae, T., Ploysri, W.: Extraction of nanocellulose from the residue of sugarcane bagasse fiber for anti-*Staphylococcus aureus* (*S. aureus*) application. Polymers **16**(11), Article ID 1612 (2024). https://doi.org/10.3390/polym16111612
39. Augaitis, N., Vaitkus, S., Członka, S., Kairytė, A.: Research of wood waste as a potential filler for loose-fill building insulation: appropriate selection and incorporation into polyurethane biocomposite foams. Materials **13**(23), Article ID 5336 (2020). https://doi.org/10.3390/ma13235336
40. Grundke, K., et al.: Experimental studies of contact angle hysteresis phenomena on polymer surfaces—toward the understanding and control of wettability for different applications. Adv. Coll. Interface. Sci. **222**, 350–376 (2015). https://doi.org/10.1016/j.cis.2014.07.007
41. Uduma, C., et al.: Synthetic modification of sunflower oil. Path of Science **8**(9), 1010–1017 (2022). https://doi.org/10.22178/pos.85-3
42. Alarcon, R., Almeida, M., Rinaldo, D., Bannach, G.: Synthesis and thermal study of polymers from soybean, sunflower, and grape seed maleinated oil. Europ. J. Lipid Sci. Technol. **119**(10), Article ID 1600515 (2017). https://doi.org/10.1002/ejlt.201600515
43. Yang, J., Ching, Y., Chuah, C., Liou, N.: Preparation and characterization of starch/empty fruit bunch-based bioplastic composites reinforced with epoxidized oils. Polymers **13**(1), Article ID 94 (2020). https://doi.org/10.3390/polym13010094

44. Cuong, T.D., Dien, L.Q., Hoang, P.H.: Preparation of bio-based porous material with high oil adsorption capacity from bio-polyurethane and sugarcane bagasse. RSC Adv. **14**(10), 6938–6947 (2024). https://doi.org/10.1039/d4ra00469h
45. Zarzyka, I.: Preparation and characterization of rigid polyurethane foams with carbamide and borate groups. Polym. Int. **65**(12), 1430–1440 (2016). https://doi.org/10.1002/pi.5198
46. Lumcharoen, D., Saravari, O.: Preparation and characterization of flexible polyurethane foams from palm oil-based polyol. Adv. Mater. Res. **911**, 352–356 (2014). https://doi.org/10.4028/www.scientific.net/AMR.911.352
47. Calcagnile, P., et al.: Magnetically driven floating foams for the removal of oil contaminants from water. ACS Nano **6**(6), 5413–5419 (2012). https://doi.org/10.1021/nn3012948
48. Tomon, T.R.B., Omisol, C.J.M., Aguinid, B.J.M., et al.: A novel naturally superoleophilic coconut oil-based foam with inherent hydrophobic properties for oil and grease sorption. Sci. Rep. **14**, Article ID 14223 (2024). https://doi.org/10.1038/s41598-024-64178-2
49. Zhu, Q., et al.: Robust superhydrophobic polyurethane sponge as a highly reusable oil-absorption material. J. Mater. Chem. A **1**(17), 5386–5393 (2013). https://doi.org/10.1039/c3ta00125c
50. Fong, M., Berrin, J.-G., Paës, G.: Investigation of the binding properties of a multi-modular GH45 cellulase using bioinspired model assemblies. Biotechnol. Biofuels **9**(1), Article ID 12 (2016). https://doi.org/10.1186/s13068-016-0428-y
51. Vanholme, R., Demedts, B., Morreel, K., Ralph, J., Boerjan, W.: Lignin biosynthesis and structure. Plant Physiol. **153**(3), 895–905 (2010). https://doi.org/10.1104/pp.110.155119
52. Kocaturk, E., Salan, T., Ozcelik, O., Alma, M.H., Candan, Z.: Recent advances in lignin-based biofuel production. Energies **16**(8), Article ID 3382 (2023). https://doi.org/10.3390/en16083382
53. Sarkanen, K.V., Ludwig, C.H.: Lignins. Occurrence, formation, structure and reactions. Wiley-Interscience, New York, NY, USA (1971)
54. Li, Y., Ragauskas, A.J.: Kraft lignin-based rigid polyurethane foam. J. Wood Chem. Technol. **32**(3), 210–224 (2012). https://doi.org/10.1080/02773813.2011.652795
55. Vieira, F.R., et al.: A semi-conductive rigid polyurethane foam derived from kraft lignin polyol and a PEDOT coating. Mater. Today Sustain. Article ID 101003 (2024). https://doi.org/10.1016/j.mtsust.2024.101003
56. Bernardini, J., Cinelli, P., Anguillesi, I., Coltelli, M.-B., Lazzeri, A.: Flexible polyurethane foams green production employing lignin or oxypropylated lignin. Eur. Polym. J. **64**, 147–156 (2015). https://doi.org/10.1016/j.eurpolymj.2014.11.039
57. Evtuguin, D.V., Andreolety, J.P., Gandini, A.: Polyurethanes based on oxygen-organosolv lignin. Eur. Polymer J. **34**(8), 1163–1169 (1998). https://doi.org/10.1016/S0014-3057(97)00245-0
58. Arshanitsa, A., Krumina, L., Telysheva, G., Dizhbite, T.: Exploring the application potential of incompletely soluble organosolv lignin as a macromonomer for polyurethane synthesis. Ind. Crops Prod. **92**, 1–12 (2016). https://doi.org/10.1016/j.indcrop.2016.07.050
59. Mimini, V., et al.: Lignosulfonate-based polyurethane materials via cyclic carbonates: preparation and characterization. Holzforschung **74**(2), 203–211 (2020). https://doi.org/10.1515/hf-2018-0298
60. Mohammadpour, R., Sadeghi, G.M.M.: Effect of liquefied lignin content on synthesis of bio-based polyurethane foam for oil adsorption application. J. Polym. Environ. **28**(3), 892–905 (2020). https://doi.org/10.1007/s10924-019-01650-5
61. Ma, X., et al.: Mechanically robust, solar-driven, and degradable lignin-based polyurethane adsorbent for efficient crude oil spill remediation. Chem. Eng. J. **415**, Article ID 128956 (2021). https://doi.org/10.1016/j.cej.2021.128956
62. Chen, J., et al.: Multifunctional superhydrophilic/underwater superoleophobic lignin-based polyurethane foam for highly efficient oil-water separation and water purification. Sep. Purif. Technol. **311**, Article ID 123284 (2023). https://doi.org/10.1016/j.seppur.2023.123284
63. Shi, Y., et al.: One-step foam assisted synthesis of robust superhydrophobic lignin-based magnetic foams for highly efficient oil-water separation and rapid solar-driven heavy oil

removal. J. Water Process Eng. **46**, Article ID 102643 (2022). https://doi.org/10.1016/j.jwpe.2022.102643
64. Wu, J., et al.: Superhydrophobic lignin-based multifunctional polyurethane foam with SiO_2 nanoparticles for efficient oil adsorption and separation. Sci. Total Environ. **860**, Article ID 160276 (2023). https://doi.org/10.1016/j.scitotenv.2022.160276
65. Rahman, M.L., et al.: Waste fiber-based poly(hydroxamic acid) ligand for toxic metals removal from industrial wastewater. Polymers **13**(9), Article ID 1486 (2021). https://doi.org/10.3390/polym13091486
66. Maimaiti, H., Arken, K., Wumaier, M., Wushuer, L.: Oil absorbency of cellulose/butylmethacrylate graft polymer fibers. Iran. Polym. J. **23**(9), 671–678 (2014). https://doi.org/10.1007/s13726-014-0261-8
67. Jarrah, K., Hisaindee, S., Al-Sayah, M.H.: Preparation of oil sorbents by solvent-free grafting of cellulose cotton fibers. Cellulose **25**(7), 4093–4106 (2018). https://doi.org/10.1007/s10570-018-1846-8
68. Li, D., Zhu, F.Z., Li, J.Y., Na, P., Wang, N.: Preparation and characterization of cellulose fibers from corn straw as natural oil sorbents. Ind. Eng. Chem. Res. **52**(1), 516–524 (2013). https://doi.org/10.1021/ie302288k
69. Paulauskiene, T., Sirtaute, E., Uebe, J.: A cellulose aerogel made from paper and hemp waste added with starch for the sorption of oil. J. Marine Sci. Eng. **11**(7), Article ID 1343 (2023) https://doi.org/10.3390/jmse11071343
70. Mai, T., Wang, P.-L., Ma, M.-G.: Promising cellulose-based aerogel composites: preparation methods and advanced applications. Compos. Part B: Eng. **281**, Article ID 111559 (2024). https://doi.org/10.1016/j.compositesb.2024.111559
71. Ottenhall, A., Seppänen, T., Ek, M.: Water-stable cellulose fiber foam with antimicrobial properties for bio-based low-density materials. Cellulose **25**(4), 2599–2613 (2018). https://doi.org/10.1007/s10570-018-1738-y
72. Jiang, S., et al.: Cellulose-based composite thermal-insulating foams toward eco-friendly, flexible and flame-retardant. Carbohydr. Polym. **273**, Article ID 118544 (2021). https://doi.org/10.1016/j.carbpol.2021.118544
73. Sun, H., et al.: Lightweight, mechanically robust and scalable cellulose-based foam enabled by organic-inorganic network and air drying. Chem. Eng. J. **491**, Article ID 152014 (2024). https://doi.org/10.1016/j.cej.2024.152014
74. Ma, S., et al.: Multifunctional hydrophobic bio-based foam toward highly efficient photothermal cleanup of high-viscosity crude oil spills, oil-water separation and aqueous organic pollutant elimination. Colloids Surf. A Physicochem. Eng. Aspects **667**, Article ID 131431 (2023). https://doi.org/10.1016/j.colsurfa.2023.131431
75. Tan, J., et al.: Synthesis of honeycomb-like carbon foam from larch sawdust as efficient absorbents for oil spills cleanup and recovery. Materials **11**(7), Article ID 1106 (2018). https://doi.org/10.3390/ma11071106
76. Liu, L., et al.: Potential of alginate fibers incorporated with drug-loaded nanocapsules as drug delivery systems. J. Mater. Chem. B **2**(43), 7596–7604 (2014). https://doi.org/10.1039/c4tb01392a
77. Lee, K.Y., Mooney, D.J.: Alginate: properties and biomedical applications. Prog. Polym. Sci. **37**(1), 106–126 (2012). https://doi.org/10.1016/j.progpolymsci.2011.06.003
78. Kuo, C.K., Ma, P.X.: Maintaining dimensions and mechanical properties of ionically crosslinked alginate hydrogel scaffolds in vitro. J. Biomed. Mater. Res., Part A **84A**(4), 899–907 (2008). https://doi.org/10.1002/jbm.a.31375
79. Adamiak, K., Sionkowska, A.: State of innovation in alginate-based materials. Marine Drugs **21**(6), 353 (2023). https://doi.org/10.3390/md21060353
80. Yang, J.-S., Xie, Y.-J., He, W.: Research progress on chemical modification of alginate: a review. Carbohyd. Polym. **84**(1), 33–39 (2011). https://doi.org/10.1016/j.carbpol.2010.11.048
81. Akshaya, S., Nathanael, A.J.: A review on hydrophobically associated alginates: approaches and applications. ACS Omega **9**(4), 4246–4262 (2024). https://doi.org/10.1021/acsomega.3c08619

82. Djemaa, I.B., et al.: One-step generation of alginate-based hydrogel foams using CO_2 for simultaneous foaming and gelation. Gels **8**(7), Article ID 444 (2022). https://doi.org/10.3390/gels8070444
83. Liu, H., Geng, B., Chen, Y., Wang, H.: Review on the aerogel-type oil sorbents derived from nanocellulose. ACS Sustain. Chem. Eng. **5**(1), 49–66 (2017). https://doi.org/10.1021/acssuschemeng.6b02301
84. Wu, M., et al.: Covalently cross-linked and hydrophobically modified alginic acid hydrogels and their application as drug carriers. Polym. Eng. Sci. **53**(8), 1583–1589 (2013). https://doi.org/10.1002/pen.23415
85. Wang, Y., Feng, Y., Yao, J.: Construction of hydrophobic alginate-based foams induced by zirconium ions for oil and organic solvent cleanup. J. Colloid Interface Sci. **533**, 182–189 (2019). https://doi.org/10.1016/j.jcis.2018.08.073
86. Pourjavadi, A., Amin, S.S., Hosseini, S.H.: Delivery of hydrophobic anticancer drugs by hydrophobically modified alginate-based magnetic nanocarrier. Ind. Eng. Chem. Res. **57**(3), 822–832 (2018). https://doi.org/10.1021/acs.iecr.7b04050
87. Cui, S., et al.: Aluminum soap nanoparticles-lignin powder form phase-selective gelator: synergistic adsorption and gelation for rapid oil spill recovery. Res. Square, preprint (2023). https://doi.org/10.21203/rs.3.rs-2718489/v1
88. Ma, X., et al.: Degradable $Ti_3C_2T_x$ MXene nanosheets containing a lignin polyurethane photothermal foam (LPUF) for rapid crude oil cleanup. ACS Appl. Nano Mater. **5**(2), 2848–2858 (2022). https://doi.org/10.1021/acsanm.1c04556
89. Wang, Y., Yu, X., Fan, W., Liu, R., Liu, Y.: Alginate–oil gelator composite foam for effective oil spill treatment. Carbohydr. Polym. **294**, Article ID 119755 (2022). https://doi.org/10.1016/j.carbpol.2022.119755
90. Balla, E., et al.: Poly(lactic acid): a versatile biobased polymer for the future with multifunctional properties—from monomer synthesis, polymerization techniques and molecular weight increase to PLA applications. Polymers **13**(11), Article ID 1822 (2021). https://doi.org/10.3390/polym13111822
91. Ren, Q., Wang, J., Zhai, W., Lee, R.E.: Fundamental influences of induced crystallization and phase separation on the foaming behavior of poly(lactic acid)/polyethylene glycol blends blown with compressed CO_2. Ind. Eng. Chem. Res. **55**(49), 12557–12568 (2016). https://doi.org/10.1021/acs.iecr.6b03266
92. Li, B., et al.: Biodegradable PLA/PBS open-cell foam fabricated by supercritical CO_2 foaming for selective oil-adsorption. Sep. Purif. Technol. **257**, Article ID 117949 (2021). https://doi.org/10.1016/j.seppur.2020.117949
93. Lima, G.M.R., Mukherjee, A., Picchioni, F., Bose, R.K.: Characterization of biodegradable polymers for porous structure: further steps toward sustainable plastics. Polymers **16**(8), Article ID 1147 (2024). https://doi.org/10.3390/polym16081147
94. Yang, H., et al.: Fabrication of bio-based biodegradable poly(lactic acid) (PLA) and poly(3-hydroxybutyrate-co-3-hydroxyvalerate) (PHBV) composite foams for highly efficient oil-water separation. Int. J. Biol. Macromol. **257**, Article ID 128750 (2023). https://doi.org/10.1016/j.ijbiomac.2023.128750
95. Xu, J., Zhang, L., Li, D., Bao, J., Wang, Z.: Foaming of poly(3-hydroxybutyrate-co-3-hydroxyvalerate) with supercritical carbon dioxide: foaming performance and crystallization behavior. ACS Omega **5**(17), 9839–9845 (2020). https://doi.org/10.1021/acsomega.9b04501
96. Kaisone, T., Harnkarnsujarit, N., Leejarkpai, T., Nampitch, T.: Mechanical and thermal properties of toughened PLA composite foams with modified coconut fiber. Appl. Mech. Mater. **851**, 179–185 (2016). https://doi.org/10.4028/www.scientific.net/AMM.851.179
97. Albuquerque, C.G., et al.: Comparative study of diesel sorption performance between Chorisia speciosa fibers and a commercial polyurethane foam. Matéria (Rio de Janeiro) **26**(1), Article ID e-1219 (2021). https://doi.org/10.1590/s1517-707620210001.1219
98. Zhang, J., et al.: A facile scalable conductive graphene-coated *Calotropis gigantea* yarn. Cellulose **29**(6), 3545–3556 (2022). https://doi.org/10.1007/s10570-022-04475-z

99. Xiao, W., et al.: Fabrication of foam-like oil sorbent from polylactic acid and *Calotropis gigantea* fiber for effective oil absorption. J. Cleaner Prod. **278**, Article ID 123507 (2021). https://doi.org/10.1016/j.jclepro.2020.123507
100. Ramesh, G., Subramanian, K., Sathiyamurthy, S., Jayabal, S.: Study on influence of fiber length and fiber weight proportion in mechanical behavior of Calotropis gigantea fiber–epoxy composites. Polym. Compos. **41**(11), 4899–4906 (2020). https://doi.org/10.1002/pc.25761
101. Montes, M.L.I., et al.: Design and characterization of PLA bilayer films containing lignin and cellulose nanostructures in combination with umbelliferone as active ingredient. Front. Chem. **7**, Article ID 157 (2019). https://doi.org/10.3389/fchem.2019.00157
102. Ding, H., et al.: High hydrophobic poly(lactic acid) foams impregnating one-step Si–F modified lignin nanoparticles for oil/organic solvents absorption. Compos. Commun. **25**, Article ID 100730 (2021). https://doi.org/10.1016/j.coco.2021.100730
103. Luo, W.-C., et al.: Impact of formulation on the quality and stability of freeze-dried nanoparticles. Eur. J. Pharm. Biopharm. **169**, 256–267 (2021). https://doi.org/10.1016/j.ejpb.2021.10.014
104. Songsaeng, S., Thamyongkit, P., Poompradub, S.: Natural rubber/reduced-graphene oxide composite materials: morphological and oil adsorption properties for treatment of oil spills. J. Adv. Res. **20**, 79–89 (2019). https://doi.org/10.1016/j.jare.2019.05.007
105. Charoeythornkhajhornchai, P., Khamloet, W., Nungjumnong, P.: Effect of carbon allotropes on foam formation, cure characteristics, mechanical and thermal properties of NRF/carbon composites. J. Cell. Plast. **58**(1), 41–57 (2022). https://doi.org/10.1177/0021955X20979548
106. Chin, C.C., Musbah, N.D., Abdullah, I., Lazim, A.M.: Characterization and evaluation of prudent liquid natural rubber-based foam for oil spill control application. Arab. J. Sci. Eng. **43**(11), 6097–6108 (2018). https://doi.org/10.1007/s13369-018-3256-5
107. Mustapa, M.H., Musbah, D.L., Sulaiman, N.M., Lazim, A.M.: Fabrication of rubber gel as oil absorbent by using water as porosity agent for oil removal. J. Appl. Polym. Sci. **136**(28), Article ID 47749 (2019). https://doi.org/10.1002/app.47749
108. Boonmahitthisud, A., Boonkerd, K.: Sustainable development of natural rubber and its environmentally friendly composites. Curr. Opin. Green Sustain. Chem. **28**, Article ID 100446 (2021). https://doi.org/10.1016/j.cogsc.2021.100446
109. Charoeythornkhajhornchai, P., Samthong, C., Somwangthanaroj, A.: Effect of graphene treated with cyclohexyl diamine by diazonium reaction on cure kinetics, mechanical, thermal, and physical properties of natural rubber/graphene nanocomposite foam. Polym. Compos. **40**(S2), E1410–E1419 (2019). https://doi.org/10.1002/pc.25160
110. Mardiyati, Y., Fauza, A.N., Rachman, O.A., Steven, S., Santosa, S.P.: A silica–lignin hybrid filler in a natural rubber foam composite as a green oil spill absorbent. Polymers **14**(14), Article ID 2930 (2022). https://doi.org/10.3390/polym14142930
111. Das, A.K., et al.: Review on tannins: extraction processes, applications and possibilities. S. Afr. J. Bot. **135**, 58–70 (2020). https://doi.org/10.1016/j.sajb.2020.08.008
112. Wei, Q., Oribayo, O., Feng, X., Rempel, G.L., Pan, Q.: Synthesis of polyurethane foams loaded with TiO_2 nanoparticles and their modification for enhanced performance in oil spill cleanup. Ind. Eng. Chem. Res. **57**(27), 8918–8926 (2018). https://doi.org/10.1021/acs.iecr.8b01037
113. Wang, C.-F., Lin, S.-J.: Robust superhydrophobic/superoleophilic sponge for effective continuous absorption and expulsion of oil pollutants from water. ACS Appl. Mater. Interfaces. **5**(18), 8861–8864 (2013). https://doi.org/10.1021/am403266v
114. Gui, X., et al.: Magnetic and highly recyclable macroporous carbon nanotubes for spilled oil sorption and separation. ACS Appl. Mater. Interfaces. **5**(12), 5845–5850 (2013). https://doi.org/10.1021/am4015007
115. Subrati, A., et al.: Developing hydrophobic graphene foam for oil spill cleanup. Ind. Eng. Chem. Res. **56**(24), 6945–6951 (2017). https://doi.org/10.1021/acs.iecr.7b00716
116. Li, X., et al.: Facile synthesis of formaldehyde-free bio-based thermoset resins for fabrication of highly efficient foams. Polymers **14**(23), Article ID 5140 (2022). https://doi.org/10.3390/polym14235140

117. Zhang, T., et al.: Design and manufacturing of cost-effective tannin-based polyurethane foam as an efficient and reusable absorbent for oil and solvents. Ind. Crops Prod. **189**, Article ID 115815 (2022). https://doi.org/10.1016/j.indcrop.2022.115815
118. Li, W., Zhang, Y., Chen, X.: Efficient liquefaction of mulberry twig ethanol-extractives for the preparation of bio-based phenolic foams. Ind. Crops Prod. **186**, 115,815 (2022). https://doi.org/10.1016/j.indcrop.2022.115815
119. Pizzi, A., Celzard, A., Tondi, G.: Chemistry, morphology, microtomography and activation of natural and carbonized tannin foams for different applications. Macromol. Symp. **313–314**(1), 100–111 (2012). https://doi.org/10.1002/masy.201250311
120. Basso, M.C., et al.: MALDI-TOF and 13C NMR analysis of tannin–furanic–polyurethane foams adapted for industrial continuous lines application. Polymers **6**(12), 2985–3004 (2014). https://doi.org/10.3390/polym6122985
121. Lopes, P.J.G., et al.: Effect of the pH value of tannin extracts on properties of classic tannin foams. J. Cell. Plast. **58**(6), 861–875 (2022). https://doi.org/10.1177/0021955X221144546
122. Liu, B., et al.: Formaldehyde-free renewable thermosetting foam based on biomass tannin with a lignin additive. J. Renew. Mater. **10**(11), 3009–3024 (2022). https://doi.org/10.32604/jrm.2022.019848

Chapter 3
Factors Affecting Foam Sorbent Performance

Abstract The effectiveness of bio-based foam sorbents in oil spill cleanups is determined by a complex interplay of factors that influence their absorption capacity, durability, and adaptability to diverse environmental conditions and oil types. This chapter delves into nine primary factors impacting the performance of these sorbents: material composition, physical properties, chemical properties, mechanical properties, thermal stability, environmental conditions, operational parameters, recyclability and reusability, and environmental impact. Each of these factors plays a significant role in enhancing the efficacy and sustainability of the sorbents under various conditions. We begin by examining the fundamental capabilities of the sorbents, focusing on how each property enhances its efficiency in different ecological settings and temperature conditions. This analysis leads to a discussion of operational parameters to shed light on the practical deployment and effectiveness of the sorbents in field conditions. The chapter then transitions to considerations of recyclability and reusability, emphasizing the sustainability of these materials in alignment with eco-friendly initiatives. Lastly, we assess the broader ecological implications of using bio-based foam sorbents in oil spill responses. By thoroughly understanding and optimizing these properties, bio-based foams can be tailored to effectively meet the challenges of real-world spill remediation, offering a sustainable alternative to synthetic options and addressing global environmental hazards.

Keywords Biopolymers · Porosity · Foam density · Chemical stability · Recyclability · Reusability · Pore size · Operational parameters · Flexibility · Thermal stability · Strength · Durability · Salinity · Biodegradability

3.1 Material Composition

The material composition of foam sorbents is crucial in determining their effectiveness in oil spill cleanups. Bio-based sorbents, typically derived from renewable sources, can be broadly categorized into natural polymers, vegetable oils, and biopolymers. These materials not only offer significant environmental benefits such

A. A. Lubguban et al., *Bio-based Foam Sorbents*, Engineering Materials,
https://doi.org/10.1007/978-981-96-1460-8_3

as biodegradability, reduced carbon footprint, and lower ecological toxicity but also pave the way for a more sustainable and optimistic future. As industries globally transition towards eco-friendly practices, integrating bio-based materials into spill response technologies becomes essential for effective environmental management [1, 2].

3.1.1 Natural Polymer

Natural polymers like polysaccharides, proteins, and lignin are commonly used in bio-based foam sorbents due to their availability and environmentally friendly nature. These materials are often derived from plants, crustaceans, or other biological sources and can be chemically modified to enhance their oil-absorbing properties [3].

Polysaccharides such as cellulose and chitin are extensively utilized in bio-based sorbents due to their robust structural configurations, which facilitate the formation of highly porous networks that enhance oil absorption capabilities. Cellulose, characterized by its numerous hydroxyl groups, can be chemically modified to increase its hydrophobicity and oil affinity, thereby improving oil uptake while minimizing water absorption. For instance, studies have shown that chemically modified cellulose foams can achieve significant oil absorption rates due to their lightweight and porous nature, which allows for high oil retention [4]. The superior oil absorption rates of cellulose-based foams can be attributed to their larger surface area and interconnected porous structures, which enhance the efficiency of oil retention [5]. Although such modifications enhance performance, they also involve additional costs that could impact the overall economic feasibility of cellulose-based sorbents. Nonetheless, cellulose's abundance and renewable nature help mitigate these costs, making it an economically viable option for large-scale applications.

Chitosan, a derivative of chitin found in crustacean shells, has emerged as a promising bio-based material. It offers excellent biocompatibility and biodegradability and can be chemically modified to improve its hydrophobicity. Research indicates that chitosan-based foams exhibit significant potential in oil absorption, showcasing the remarkable advancements in sustainable materials. For example, due to their favorable surface characteristics, chitosan has been effectively utilized in developing bio-sorbents that demonstrate high adsorption capacities for various pollutants, including oils [6]. Furthermore, incorporating metal ions, such as zinc, into chitosan matrices has been shown to enhance the adsorption performance of these materials, making them even more effective in oil spill remediation applications.

Moreover, the versatility of polysaccharides in modifying their chemical and physical properties allows for the design of tailored sorbents that can effectively address the complexities of oil spill cleanup. Combining cellulose and chitosan in hybrid sorbent formulations has been explored to leverage the unique properties of both materials. Such hybrid systems can provide improved mechanical strength and enhanced oil absorption capabilities, making them suitable for challenging environmental conditions [4, 5].

Proteins such as keratin and casein are valuable in sorbent manufacturing due to their functional groups, which interact with oil molecules through various bonding mechanisms, including hydrogen bonding and hydrophobic interactions. These interactions enhance the selectivity of sorbents for oil over water, which is crucial in environments where oil is mixed with seawater or freshwater. A study by Ifelebuegu and Johnson [7] highlights the potential of keratin-based sorbents, noting their effectiveness in oil/water separation and spill cleanup applications due to their unique chemical properties and structural integrity. The authors emphasize that keratin, derived from poultry feathers, is abundant and can be processed into effective sorbents with high oil sorption capacities.

Research conducted by Kelle and Eboatu [8] also demonstrates that chicken feathers, rich in keratin, can serve as effective sorbents for crude oil and its lower fractions. The study indicates that the porous structure of keratin fibers allows for significant oil absorption, making them suitable for real-world oil spill scenarios. The inherent hydrophobicity of keratin and its ability to form strong interactions with oil molecules contribute to its effectiveness as a sorbent material.

Casein, another protein with promising sorption properties, has been studied for its ability to enhance oil absorption in composite materials. Although the specific study by Du et al. [9] focuses on collagen fibers rather than casein, it suggests that protein matrices can significantly improve the oil sorption capacity of composites, demonstrating the importance of protein-based materials in foam sorbents. The study indicates that the protein's functional groups facilitate strong interactions with oil molecules, enhancing the overall performance of the sorbent.

In addition to their oil absorption capabilities, protein-based materials in foam sorbents contribute to the overall strength and durability of the sorbents, enabling them to withstand mechanical stresses during deployment and recovery. For example, the work of Singh et al. [10] highlights the development of a novel natural sorbent using unprocessed raw cotton, which exhibits enhanced mechanical properties and oil sorption capacity when combined with protein-based materials. This synergy between natural fibers and proteins allows for creation of robust sorbents that can effectively address the challenges posed by oil spills.

Lignin, a paper and pulping industry by-product, offers significant potential as a bio-based foam material. Naturally hydrophobic and oleophilic, lignin is well-suited for oil spill cleanups [11]. In addition to its superior oil absorption capabilities, lignin-based foams exhibit high thermal stability and mechanical strength [12]. These properties make lignin a viable competitor to synthetic foams while maintaining the environmental advantage of biodegradability [13]. For instance, Alinejad et al. [11] discuss the incorporation of lignin into polyurethane products, highlighting its potential to create flexible and rigid foams that can effectively absorb oils. Furthermore, research by Luo et al. [12] demonstrates that lignin can serve as a bio-polyol in the production of rigid, lightweight composite foams, showcasing its effectiveness in enhancing the mechanical properties of the resulting materials.

Moreover, studies have shown that lignin can be modified to improve its compatibility with other materials, thus enhancing its performance in foam applications. For example, the work by Quan et al. [14] focuses on recovering kraft lignin directly from black liquor, which can then be utilized in flexible polyurethane foams, demonstrating lignin's versatility and effectiveness in various foam formulations. Additionally, the valorization of lignin aligns with a circular economy approach by utilizing waste products, making it both a sustainable and cost-effective option [15]. The potential for lignin to replace petroleum-based polyols in polyurethane formulations reduces reliance on fossil fuels and contributes to the development of environmentally friendly materials [16].

Having explored the multifaceted role of natural polymers like cellulose, chitosan, and lignin in enhancing oil spill cleanup efficiency, we now turn to another category of bio-based materials: vegetable oils. Like natural polymers, vegetable oils contribute unique properties to foam sorbents but focus more on flexibility and durability, key factors in their performance during oil spill remediation.

3.1.2 Vegetable Oils

Vegetable oils, such as soybean, castor, and coconut oil, are increasingly used to create bio-based foams. These oils can be polymerized to form flexible, oleophilic foams with excellent absorption properties.

Soybean oil, one of the most widely available vegetable oils, can be polymerized into bio-based foams that are both effective and sustainable. The high triglyceride content in soybean oil produces foams with excellent structural properties, enabling them to absorb significant amounts of oil [17]. Additionally, soybean-based foams can be modified to improve durability and resistance to mechanical stress, making them more suitable for large-scale cleanup efforts. A study by Li et al. [18] highlights that bio-polyurethane foams derived from soybean oil exhibit superior biodegradability compared to petroleum-based counterparts, reinforcing the sustainability aspect of using soybean oil in foam production. Despite the inherent challenges in ensuring long-term stability, the abundance and renewability of soybean oil make it an attractive option for producing bio-based foams that align with sustainability principles.

Similarly, castor oil offers unique advantages in bio-based foam production due to its high hydroxyl content, which allows for enhanced flexibility and oil absorption capabilities. Castor oil-based foams are particularly valued for maintaining mechanical strength while offering high oil sorption capacities. Research by Purwanto et al. [19] demonstrates that castor oil can effectively create non-isocyanate polyurethane foams, which exhibit excellent mechanical properties and efficiency in oil absorption. This balance of flexibility and performance makes castor oil an ideal material for bio-based foams designed for repeated use in oil spill remediation. Moreover, castor oil's natural viscosity and chemical structure lend themselves to resilient foams that can maintain their absorption efficiency over multiple use cycles.

Coconut oil, known for its inherent hydrophobicity, has also proven to be an excellent foundation for bio-based foams with enhanced oleophilic properties. Recent research by Tomon et al. [20] has demonstrated the potential of coconut oil-based polyols to create foams that absorb oil efficiently and possess strong recyclability and chemical stability. These foams are highly effective in absorbing oil while repelling water, thanks to coconut oil's natural saturation and fatty acid composition. The development of coconut oil-based foams further underscores the potential of vegetable oils in advancing sustainable materials for oil spill cleanups, offering both environmental benefits and practical performance advantages. By utilizing renewable, plant-based oils, these bio-based foams represent a step forward in creating greener, more sustainable sorbents for real-world environmental applications [21].

As we delve deeper into the spectrum of bio-based materials, the transition from vegetable oils such as soybean, castor, and coconut oil, which primarily enhance oleophilicity and flexibility, to biopolymers introduces a shift towards addressing not just the physical and chemical robustness but also the environmental sustainability of sorbents. Biopolymers like PLA and PHA exemplify the next step in optimizing foam sorbents' lifecycle and functional adaptability in challenging environmental conditions.

3.1.3 Biopolymers

Biopolymers such as polylactic acid (PLA) and polyhydroxyalkanoates (PHA) are utilized for their biodegradability and mechanical stability, providing a durable backbone for sorbents that can endure the physical and chemical rigors of oil spill environments [22]. Advances in biopolymer technology have led to the development of sorbents that not only effectively absorb oil but also degrade over time, thus reducing the long-term environmental footprint of cleanup operations. For example, Wang et al. [22] demonstrated that hybrid electrospun porous fibers made from PLA and nano-ZIF-8@C600 exhibited excellent oil absorption capabilities while maintaining biodegradability, making them suitable for environmental applications. Their study highlights the potential of PLA as a promising environmentally friendly material for producing effective sorbent materials.

Moreover, research by Alassod et al. [23] explored the use of polypropylene/lignin blend monoliths as sorbents in oil spill cleanup, emphasizing the role of lignin in enhancing the absorption capacity of the resulting materials. This study illustrates the versatility of biopolymers in creating effective sorbents that can withstand harsh conditions while providing environmental benefits. Similarly, the work of Bhagwat and Jaspal [24] emphasizes the need for biodegradable materials in oil spill cleanup, noting that various synthetic polymers have limitations regarding biodegradability and non-renewability. Their findings support the idea that biopolymers like PLA and PHA can serve as sustainable alternatives in the development of sorbents.

Furthermore, incorporating biopolymers into composite materials has shown promise in enhancing oil sorption performance. For instance, the study by Zeng et al. [25] on biopolymer membranes for oil–water separation demonstrated that these materials possess unique wettability properties that contribute to their effectiveness in oil spill remediation. The biocompatibility and biodegradability of these membranes further align with the principles of sustainable material development.

Moving from synthetic and semi-synthetic biopolymers to those derived through microbial processes, we encounter materials such as xanthan and alginate. These microbial biopolymers continue to uphold the principles of biodegradability and mechanical stability and offer tailored properties that can be engineered to target the nuances of various oil spill conditions specifically.

Microbial biopolymers, particularly xanthan and alginate, are instrumental in remediating oil spills. These biopolymers, produced by microbial processes, are highly valued for their customizable properties, which can be tailored to target specific oil types [26]. This capability makes them crucial in developing specialized sorbents that can adapt to different spill conditions. For instance, xanthan gum, a polysaccharide produced by the bacterium *Xanthomonas campestris*, has been shown to enhance the viscosity of oil dispersions, thereby improving the effectiveness of oil recovery methods [27]. Research by Balíková et al. [28] highlights the potential of xanthan in removing heavy metals from contaminated water, demonstrating its versatility and effectiveness in environmental remediation.

Similarly, alginate, derived from brown seaweed, has been extensively studied for its oil absorption properties. A study by Jung and Hu [29] demonstrated that sodium alginate can be effectively used in the remediation of oil-contaminated sand, showcasing its ability to encapsulate oil and facilitate its removal from the environment. The authors noted that alginate's gel-forming ability allows for the creation of sorbents that can effectively capture and retain oil, making it a valuable material in spill response strategies.

The versatility of microbial biopolymers extends further as they can be engineered to vary in degradation rates, offering additional control over the lifecycle and environmental impact of the sorbents. For example, Sharma et al. [26] reviewed the applications of microbial biopolymers in ecological remediation, noting that the ability to tailor the degradation rates of these materials can enhance their effectiveness in specific contexts, such as oil spill cleanup. This adaptability allows for the development of sorbents that absorb oil and degrade in a controlled manner, minimizing long-term environmental impacts.

Moreover, incorporating biopolymers into composite materials has shown promise in enhancing oil sorption performance. Research by Cao et al. [27] indicated that biopolymers like xanthan and alginate can improve the mobility and effectiveness of oil recovery processes by modifying the physical properties of the sorbents. Their findings suggest that biopolymers can lead to more efficient oil spill remediation techniques, further supporting the role of microbial biopolymers in sustainable environmental management.

As we transition from discussing the diverse material composition of bio-based foam sorbents to exploring their physical properties in the subsequent section, it becomes crucial to understand how the intrinsic characteristics of bio-based materials translate into tangible, physical advantages in oil spill scenarios. The inherent qualities of these materials not only dictate their environmental impact but also influence their physical properties and overall effectiveness in harsh conditions. Thus, Sect. 3.2 will delve deeper into how these physical properties, shaped by the material composition, contribute to the performance and durability of foam sorbents under operational stresses, ensuring that they meet the practical demands of oil spill cleanups while aligning with sustainable practices.

3.2 Physical Properties

The effectiveness of bio-based foam sorbents in oil spill recovery is significantly influenced by their physical properties. This section examines critical characteristics such as porosity, surface area, pore size distribution, density, texture, and flexibility, elucidating how these properties impact the functionality and overall performance of sorbents in aquatic environments. Physical properties are the tangible attributes that determine how a sorbent interacts with oil and water. Each of these characteristics plays a pivotal role in the efficiency of the sorbent, affecting everything from oil absorption rate to durability and deployability in field conditions.

3.2.1 Porosity, Surface Area, and Pore Size Distribution

Porosity, defined as the measure of void spaces within a material, critically determines a foam sorbent's capacity to absorb oil. High porosity allows more space for oil to infiltrate and be trapped, thereby enhancing the absorption capacity. Bio-based foams, characterized by their high porosity, have demonstrated superior oil absorption properties due to their extensive internal networks that facilitate deeper penetration of oil into the foam structure. Wu et al. [30] found that the absorption capacity of polymer foams is significantly influenced by their density and porosity, suggesting that careful manipulation of these parameters can enhance performance in oil absorption applications. This finding underscores the need for precise engineering in foam production to optimize environmental and operational efficiency.

However, balancing porosity with structural integrity is crucial, as excessive porosity may compromise mechanical strength and limit reusability. Complementing porosity, the surface area provides additional contact points for oil molecules, facilitating quicker and more efficient absorption. This characteristic significantly influences the capacity of foam sorbents to interact with and adsorb oil. Recent studies, including that of Hoang et al. [31], have shown that bio-based sorbents with larger surface areas exhibit superior oil absorption characteristics, thanks to a bio-polyurethane foam using a bio-polyol derived from vegetable oil, which demonstrated high oil adsorption capacity of up to 14.89 g of oil per gram of foam due to its increased surface area and porosity. These results highlight the potential of bio-based materials to provide effective solutions for oil spill remediation, emphasizing their advantage over traditional synthetic foams in environmental responsiveness.

The relationship between surface area and oil absorption is further supported by the work of Than et al. [32], who explored the characteristics of palm kernel oil bio-polyol modified polyurethane foams. Their findings indicated that the introduction of bio-polyols improved the foam's structural characteristics, essential for enhancing oil uptake. This suggests that the careful design of the foam's microstructure can lead to improved performance in oil adsorption applications, reinforcing the importance of innovative material design in enhancing environmental sustainability.

Moreover, the microstructural design of bio-based foams is crucial for achieving a finely distributed network of pores. Song et al. [33] demonstrated that heat-induced bubble expansion could significantly increase the porosity of bio-based macroporous polymers, resulting in enhanced oil absorption capabilities. By optimizing the foam's microstructure, researchers can create materials that effectively remove even thin layers of oil from water surfaces, which is vital for environmental cleanup efforts. These advances indicate a shift towards more sustainable sorbent technologies that not only meet but exceed the performance standards set by traditional synthetic options.

Effective oil absorption also depends on the distribution of pore sizes within the foam. A balanced mix of small and medium-sized pores ensures rapid infiltration and secure retention of oil, preventing leakage back into the environment. Larger pores contribute to the overall absorption capacity but can influence the rate of oil drainage. Optimizing pore size distribution involves engineering foams that utilize small pores for capillary action to enhance retention, while larger pores facilitate quicker uptake. This is supported by studies like Zhu et al. [34] and Xiao et al. [35], which found that specific foam designs demonstrated exceptional oil/water selectivity due to optimized pore structures. These findings are echoed in another study by Alassod et al. [23], which explored polypropylene/lignin blend sponges exhibiting favorable properties for oil spill cleanup. The researchers noted that the pore structure of these sponges played a vital role in their oil absorption capabilities, emphasizing the need for a balanced pore size distribution to enhance both uptake and retention.

A practical illustration of the impact of optimized bio-based foams can be seen in the case study conducted by Mardiyati et al. [36], where a silica-lignin hybrid filler was incorporated into a natural rubber foam composite. This innovative approach resulted in a green oil spill absorbent that exhibited excellent oil absorption properties, demonstrating the effectiveness of bio-based materials in real-world applications. The study highlighted that the combination of lignin, a renewable resource, with silica enhanced the foam's porosity, surface area, and pore size distribution, leading to improved oil retention capabilities.

Further research underscores the potential of engineered bio-based foams with increased porosity, optimized surface area, and strategic pore size distribution, crucial for prolonged recovery operations in oil spill remediation. Studies like those by Liu et al. [37] and Lorevice et al. [38] have shown that aerogel-type sorbents from nanocellulose and cellulose nanofibril-natural rubber latex composite foams, respectively, provide superior oil absorption due to their unique porous structures and large surface areas. These enhancements allow them to outperform traditional synthetic sorbents, particularly when modified to increase hydrophobic properties, thereby improving oil retention capabilities essential for effective remediation.

Further advancements in the field are evidenced by works like those of Kurańska et al. [39] and Gu et al. [40], who have utilized biopolymers and hybrid materials to improve the compressive strength and ecological efficiency of bio-based foams. The combined findings from these studies reveal that bio-based foams with carefully engineered porosity and pore size distribution offer effective and sustainable alternatives to conventional oil spill sorbents, providing robust solutions for a variety of environmental conditions.

Understanding the interaction between porosity, surface area, and pore size distribution facilitates the design of bio-based foams that are not only effective in absorbing oil but are also durable and environmentally friendly. Incorporating sustainable practices into the material design process ensures that these sorbents can be used effectively across various spill conditions without exacerbating environmental damage.

3.2.2 Density and Buoyancy

The density of a sorbent is a critical factor influencing its buoyancy and, consequently, its effectiveness in aquatic environments, particularly during oil spill recovery operations. Sorbents must possess sufficient buoyancy to remain at the water's surface, where oil tends to accumulate. Materials with low density are favored for this purpose, as they are lightweight and capable of floating, thus ensuring prolonged contact with surface oil and enhancing absorption efficiency [41, 42]. Research indicates that the structural characteristics of sorbents, such as porosity and density, directly affect their oil absorption capabilities. Lower-density foams typically exhibit a more open structure, facilitating oil penetration into their void spaces, which is essential for effective sorption [7, 43].

Bio-based sorbents, such as those derived from expanded polysaccharides or low-density polyurethane foams, exemplify the engineering of materials to optimize both density and buoyancy. For instance, studies have demonstrated that polyurethane foams modified with various additives can significantly enhance their oil absorption capacity while maintaining buoyancy [42, 44]. This careful balance is crucial, as it ensures that the sorbents can effectively absorb oil without sinking, even after becoming saturated. The optimization of density not only enhances buoyancy but also contributes to the overall structural integrity of the sorbent, which is vital for repeated use in harsh marine conditions [41, 45].

However, it is important to note that while lower-density foams generally provide higher porosity and buoyancy, excessively low densities may compromise the durability of the sorbents under mechanical stress, potentially diminishing their effectiveness in real-world applications [7]. Therefore, achieving an optimal density is essential to maintain a balance between high oil absorption capacity and the necessary structural integrity for enduring performance in challenging environments [42, 43].

Recent studies have highlighted the performance of various natural and synthetic sorbents in oil spill scenarios. For example, Liu et al. [46] assessed the risk of oil spills in the Bohai Sea and emphasized the need for effective recovery equipment, such as sorbents, to mitigate environmental impacts. Additionally, research by Vasilyeva et al. [47] evaluated the efficiency of commercial sorbents in Arctic waters, providing empirical data on their sorption capacities under varying conditions. These findings underscore the importance of selecting appropriate sorbent materials based on their physical properties, such as density and buoyancy, to optimize their performance in oil spill remediation efforts.

The interplay between density, buoyancy, and absorption capacity is crucial in the design and application of sorbents for oil spill recovery. By engineering bio-based sorbents that balance lightweight buoyancy with robust durability, we not only enhance their performance but also bolster their environmental sustainability. Integrating these materials into spill response strategies addresses immediate oil recovery needs while fostering long-term ecological protection. This underscores the pivotal role of innovative material science in developing solutions that are operationally effective and environmentally responsible.

3.2.3 Texture and Flexibility

The texture and flexibility of sorbents critically influence their deployability and effectiveness in diverse environmental conditions. Flexible sorbents, capable of covering extensive areas and conforming to irregular surfaces, are crucial in various spill contexts. The interaction between oil and the sorbent's surface texture—whether rough or smooth—affects oil adhesion and absorption efficiency. Studies underscore that surface morphology and flexibility are directly correlated with oil recovery efficiency [1, 44, 48].

Bio-based sorbents, such as those derived from keratin and alginate, showcase an optimal blend of fibrous textures and significant flexibility, enhancing their practicality. These materials facilitate efficient deployment across water surfaces and around spills, adapting seamlessly to environmental and operational demands. Their biodegradability and performance retention over multiple uses highlight their contribution to sustainable cleanup practices [1, 49, 50]. The potential for reusability with minimal efficiency loss underscores their design's alignment with environmental sustainability goals [44, 51].

Material modifications in bio-based foam sorbents significantly enhance their texture and flexibility, which in turn improves oil adhesion and effectiveness in challenging environmental conditions. For example, research by Perera et al. [52] demonstrated how enhancements to castor oil-based polyurethane foams with octadecylsilane and diatomaceous earth substantially increased the foam's oil absorption capacity by 2.91 times compared to its unmodified counterpart. This inclusion not only improved the foam's texture for better oil adherence but also its flexibility, facilitating easier deployment across varied surfaces during oil spill cleanups. Similarly, a study by Malewska et al. [53] explored the effects of different concentrations of rapeseed-oil-based polyol and water on the structural and mechanical properties of flexible polyurethane foams. It was found that higher water content led to foams that were less dense and softer yet more resilient, making them ideally suited for use in dynamic marine environments where they need to adapt to irregular surfaces to effectively adhere to oil.

Furthermore, Coste et al. [54] explored the synthesis of bio-based soft foams through the aza-Michael reaction to replicate the open-cell structure of traditional polyurethane foams. These foams exhibited superior flexibility and texture, crucial for effective deployment in challenging conditions, with the open-cell structure enhancing oil penetration and retention, thereby improving the overall performance of the sorbent in oil spill scenarios. Additionally, research by Furtwengler et al. [55] on novel rigid polyisocyanurate foams derived from synthesized bio-based polyester polyols highlighted how structural modifications during the synthesis process could enhance mechanical properties. These foams maintained necessary flexibility and provided the strength needed to withstand the rigors of deployment in marine environments, enabling them to adapt to various surface conditions while retaining structural integrity for effective oil spill remediation.

Recent advancements in material science have led to the development of hybrid sorbents and other innovative composites that enhance oil absorption capabilities. These materials not only meet immediate recovery needs but also adhere to principles of ecological responsibility, emphasizing the ongoing evolution of sorbent technology towards sustainability [56, 57]. The integration of advanced material science in sorbent design is pivotal, focusing not only on immediate efficacy but also on long-term environmental impact.

The strategic design of bio-based sorbents, with an emphasis on optimizing texture and flexibility, is crucial for enhancing their deployability and effectiveness in oil spill recovery. These materials not only adapt to diverse environmental conditions but also support sustainable management practices by remaining functional and reusable across multiple spill events. Through advancements in material science, the development of sorbents that are both operationally effective and environmentally friendly represents a significant step forward in our ability to manage ecological challenges effectively. This approach underscores the ongoing innovation in sorbent technology, prioritizing both immediate efficacy and long-term environmental sustainability.

Having explored the vital physical properties of bio-based foam sorbents in Sect. 3.2, such as porosity, surface area, and density, which underpin their functionality and efficiency in oil spill scenarios, we now turn our attention to the equally critical chemical properties that influence sorbent performance. Section 3.3 will delve into how the chemical composition and interactions of these materials not only affect their ability to absorb and retain oil but also their resistance to environmental degradation and reusability. By examining the chemical behaviors of bio-based materials, such as their hydrophobicity, oleophilicity, and biodegradability, we can gain a fuller understanding of their long-term effectiveness and ecological impact. This transition from physical to chemical properties will provide a comprehensive insight into the design and optimization of sorbents that are both highly effective in oil recovery and aligned with environmental sustainability goals.

3.3 Chemical Properties

The effectiveness of bio-based foam sorbents in oil spill recovery is significantly influenced by their chemical properties. This section examines critical characteristics such as chemical stability, compatibility with various oil types, and the potential for chemical modifications, elucidating how these properties impact the functionality and overall performance of sorbents in aquatic environments. Chemical properties are the inherent attributes that determine how a sorbent interacts with different pollutants and environmental conditions. Each of these characteristics plays a pivotal role in the efficiency of the sorbent, affecting everything from oil absorption efficiency to environmental safety and adaptability in field conditions.

3.3.1 Chemical Stability

The chemical stability of bio-based foam sorbents for oil spill recovery is a critical area of research, particularly in understanding how their chemical composition influences resilience against environmental factors such as seawater, UV radiation, and temperature fluctuations. Recent studies have highlighted the importance of both natural and engineered materials in maintaining their integrity and functionality over time.

Bio-based foams, such as those derived from cellulose and natural rubber latex, have been shown to exhibit significant resilience when exposed to harsh environmental conditions. For instance, Lorevice et al. [38] demonstrated that porous cellulose nanofibril-natural rubber latex composite foams possess excellent absorption capabilities for oil and organic solvents, indicating their potential for effective oil spill remediation. The inherent chemical properties of these materials, such as their biodegradability and low density, contribute to their performance in diverse conditions. Additionally, the stability of these foams can be enhanced through the incorporation of nanoparticles, which have been shown to improve foam stability by increasing film thickness and electrostatic interactions, thereby reducing breakage rates under varying environmental conditions [58].

The durability of bio-based polyurethane foams has also been a subject of investigation. King et al. [59] examined the performance attributes of shape memory poly(β-hydroxythioether) foams in environmental applications, demonstrating that their ultralight structure and thermal responsiveness are critical for effective use in prolonged remediation scenarios. The study emphasized that the chemical composition of the foam plays a vital role in its resistance to hydrolytic and oxidative degradation, which are common challenges in marine environments. Furthermore, while

Wang et al. [60] focused on foam-like composites for CO2 capture, their findings on mechanical stability and durability can be extrapolated to suggest that similar properties are essential for maintaining efficient oil sorption capabilities in bio-based foams when subjected to fluctuating temperatures and exposure to seawater. The study by Xu et al. [61] on edible egg yolk peptide nanoparticles also indicated that the foaming properties of protein-based materials are significantly influenced by environmental conditions such as pH and temperature, which can affect their stability and performance.

Overall, the chemical composition of bio-based foam sorbents significantly influences their stability and effectiveness in oil spill recovery operations. The incorporation of natural materials, engineered composites, and nanoparticles can enhance their resilience against environmental challenges, ensuring that they maintain their integrity and functionality over time. This understanding is crucial for developing effective strategies for oil spill remediation in diverse environmental conditions.

3.3.2 Compatibility with Pollutants

The interaction between sorbents and pollutants is primarily governed by their chemical affinity towards oil relative to water. Delving into the hydrophobic and oleophilic properties of bio-based foam sorbents, these materials are specifically engineered to selectively absorb oil while repelling water. The effectiveness of various sorbent materials in dealing with different types of oil—from light petroleum products to heavy crude oils—is critical in determining their utility in oil spill recovery. Recent research has illuminated how the chemical composition and surface characteristics of these materials enhance their affinity for oil over water, thereby boosting their efficiency in diverse spill conditions.

Natural sorbents such as those derived from water spangle and milkweed fibers have shown exceptional oil absorption capabilities. For example, Kathir and Carroll [2] highlights that the superhydrophobic trichome-covered surfaces of water spangle leaves enable efficient oil absorption. Similarly, milkweed fibers are noted for their high oil absorption capacity due to their hydrophobic-lyophilic properties. These findings underscore the importance of surface morphology in enhancing the sorbent's chemical compatibility with oil, crucial for effective remediation efforts.

Moreover, the use of expanded polystyrene (EPS) waste for crude oil cleanup has been explored by Naser [62], demonstrating significant oleophilicity suitable for recovering oil from water surfaces. This study exemplifies the practical application of sorbent materials in real-world scenarios, often used by hand crews to manage spills that are not accessible by mechanical means.

Advancements in polymer chemistry have also contributed to the development of oleophilic gel polymers, as studied by Ali [63]. These hydrophobic acrylate polymers absorb oil by allowing it to penetrate their pore spaces, demonstrating the essential role of chemical compatibility in oil sorption. Anwana and Ewemoje [64] further confirm the hydrophobic and oleophilic characteristics of natural fibers like kapok and coir under both static and dynamic conditions, enhancing our understanding of sorbent behavior in varied environments.

Subramoniapillai and Thilagavathi [65] reviewed the performance of natural fibers in oil spill cleanups, emphasizing that ideal sorbents should not only possess excellent hydrophobic-oleophilic properties but also exhibit rapid sorption kinetics and maintain good buoyancy. This comprehensive analysis supports the need for sorbents that can effectively adapt to different oil types and spill conditions, a theme further explored by Kim et al. [48]. Their comparative study across various oil types highlights the variability in sorbent efficiency, stressing the necessity for tailored materials to cope with the specific challenges posed by different oils.

The hydrophobic and oleophilic properties of bio-based foam sorbents play a pivotal role in their effectiveness in oil spill recovery. The integration of natural fibers and engineered materials into sorbent designs shows promising avenues for enhancing oil recovery efficiency in diverse environmental conditions, as demonstrated by recent case studies and experimental data.

3.3.3 Degradation and Leaching

While bio-based sorbents offer many environmental benefits, it is essential to assess the potential for chemical degradation and the leaching of harmful components into aquatic systems. The exploration of the biodegradability of sorbent materials and their potential to release substances that could impact marine ecosystems is vital. This analysis helps identify strategies to mitigate these risks, underscoring the importance of using environmentally benign materials and incorporating safe degradation pathways to highlight advancements in sustainable sorbent technologies.

A significant study by Li et al. [18] demonstrated that bio-polyurethane (BPU) foams produced from agricultural wastes exhibit superior biodegradability compared to traditional petroleum-based foams. This research indicated that the biodegradation rate of BPU foams increases with the bio-oil content, with notable weight loss observed by week eight of the study. This finding suggests that the incorporation of bio-based materials enhances the foam's ability to degrade in natural environments, potentially reducing the risk of harmful substances leaching into aquatic ecosystems.

Research by Kurańska et al. [21] (previously mentioned in previous sections) explored the implementation of circular economy principles in the synthesis of polyurethane foams, highlighting the importance of using bio-based polyols derived from renewable resources. Their findings indicated that foams synthesized from waste materials not only reduce reliance on petrochemicals but also exhibit improved thermal stability and lower environmental impact. This aligns with the growing emphasis on sustainable materials that minimize ecological footprints.

As discussed in Sect. 3.1, Tomon et al. [20] introduced a novel coconut oil-based foam with inherent hydrophobic properties, which demonstrated effective oil and grease sorption. While the study primarily focused on the foam's sorption capabilities, it also raised concerns about the potential leaching of substances during degradation, emphasizing the need for thorough assessments of environmental safety. This highlights the dual focus on performance and safety in the development of bio-based sorbents.

Research by Ranganathan et al. [66] on biomass thermoplastic copolyamide elastomers indicated that these materials could be synthesized with a focus on sustainability, potentially leading to foams that degrade safely in marine environments. The study suggests that the careful selection of raw materials can mitigate the risks of toxic substance release during degradation.

Moreover, Huang et al. [67] conducted a thermal stability analysis of polyurethane foams made from microwave liquefaction bio-polyols, revealing that these materials exhibit favorable degradation properties. Their findings suggest that bio-based foams can be engineered to minimize harmful emissions during degradation, thus protecting marine life.

To further address environmental concerns, strategies such as the use of environmentally benign materials and the incorporation of safe degradation pathways are being developed. For instance, Valencia et al. [68] discussed the potential of bio-based foams with ultrahigh zeolite loadings for selective capture of carbon dioxide, highlighting their low cost and biodegradability as significant advantages. This approach not only enhances the performance of sorbents but also aligns with sustainability goals.

These studies collectively underscore the importance of biodegradability and environmental safety in the development of bio-based foam sorbents for oil spill recovery. The incorporation of renewable resources and the focus on minimizing harmful leaching during degradation are critical for ensuring the safety of these materials in marine environments.

3.3.4 Modification Techniques

Enhancing the chemical properties of sorbents through modification techniques is a critical area of development in sorbent technology. The employment of various chemical modifications, such as cross-linking and surface functionalization, has been shown to significantly improve the sorbents' oil affinity and capacity. These modifications also take into account the environmental implications of such enhancements, ensuring that the development aligns with sustainability principles.

A notable approach includes the use of modified fillers in polyurethane (PU) foams. Research by Członka et al. [69] demonstrated that incorporating walnut shell fillers treated with perlite, montmorillonite, and halloysite into PU foams results in improved mechanical properties and oil sorption capacity. The modifications enhance the morphology and viscosity of the foams, leading to better performance in oil recovery applications. This study highlights how the selection and treatment of fillers can significantly influence the effectiveness of foam sorbents.

Kairytė et al. [70] explored vacuum-based impregnation of liquid glass into sunflower press cake particles, resulting in bio-based rigid PU foams with enhanced properties. These modified foams exhibited comparable density and improved oil affinity, suggesting that chemical modifications can effectively tailor foam sorbents for specific oil spill scenarios. Additionally, Abdullah et al. [71] examined the effect of chicken feathers on the sorption properties of PU foam. The study found that the modified foam demonstrated improved oil absorption capabilities, indicating that natural organic materials can effectively enhance the oleophilic properties of foam sorbents, thereby increasing their efficiency in oil spill recovery.

Surface functionalization techniques have also shown promise in enhancing the hydrophobic and oleophilic characteristics of foam sorbents. Zimmermann et al. [41] developed a sorbent system based on organosilane-coated PU foam, which exhibited improved oil sorption capacity due to the enhanced hydrophobicity imparted by the silane treatment. This modification not only improved the performance of the foam but also utilized environmentally benign materials, aligning with sustainability principles.

However, there are concerns about the potential leaching of harmful substances from modified foam sorbents. García-Rengifo et al. [72] highlighted that while chemical modifications can enhance performance, they may also increase the environmental burden associated with the production of certain additives, such as acetic anhydride used in acetylation processes. This underscores the importance of evaluating the environmental implications of chemical modifications to ensure that the benefits of enhanced performance do not come at the cost of ecological safety.

To mitigate these risks, sustainable modification strategies are being developed. For example, Członka et al. [73] demonstrated that the use of biopolyols derived from renewable resources improves the properties of PU foams while reducing their environmental impact. This approach not only enhances the performance of foam sorbents but also promotes the principles of a circular economy by utilizing waste materials.

These advancements in bio-based foam sorbent technology through chemical modification techniques have shown significant potential for improving oil affinity and capacity. The careful selection of materials and modification strategies enhances the performance of foam sorbents while addressing environmental concerns, making it essential for continued research in this area to develop effective and sustainable solutions for oil spill recovery.

3.4 Mechanical Properties

The performance and longevity of bio-based foam sorbents in oil spill remediation are profoundly affected by their mechanical properties. This section explores essential attributes such as compressive strength, tensile strength, and elasticity, discussing how these properties influence the sorbents' functionality and durability in aquatic environments. Mechanical properties are the foundational qualities that dictate how a sorbent withstands physical stresses and environmental conditions. Each characteristic is crucial for the sorbent's operational effectiveness, impacting everything from its ability to resist deformation to its durability and reusability in challenging field conditions.

3.4.1 Introduction to Mechanical Stress

During the deployment and recovery of oil spill sorbents, the materials are subjected to various forms of mechanical stress, including compression, tension, and shear. These stresses can significantly affect the sorbents' ability to perform effectively over multiple uses. Mechanical stress tests, which simulate these real-world conditions, are critical for assessing the longevity and robustness of sorbents. By understanding and quantifying the impact of such stresses, manufacturers can design sorbents that withstand physical degradation and continue to perform optimally in diverse conditions.

Recent studies have provided valuable insights into how mechanical stresses influence the performance and longevity of bio-based foam sorbents used for oil spill recovery. Carranza et al. [74] characterized the mechanical behavior of polymeric foams under complex loading conditions, including uniaxial compression and tension, highlighting how the mechanical properties, such as yield strength and energy absorption capacity, are significantly affected by the type of loading applied. Bolf et al. [75] evaluated the shear properties of lightweight PVC core foams, revealing that specimens loaded in the tensile direction exhibited lower shear modulus values compared to those subjected to compressive loading. This suggests that foam sorbents designed for oil spill recovery must be capable of withstanding both tensile and compressive stresses, as these conditions can occur during deployment in turbulent water or when recovering oil from surfaces.

Fathi et al. [76] investigated the compressive response and energy absorption characteristics of foam-filled aluminum honeycomb composites. The study demonstrated that the mechanical response of foams can be tailored through material selection and structural design, crucial for enhancing the durability of foam sorbents in oil spill applications. This ability to absorb energy effectively can prevent structural failure during use, thereby extending the lifespan of the sorbents. Mazzuca et al. [77] examined the mechanical behavior of polyurethane foams at elevated temperatures, revealing how temperature variations can significantly impact the compressive and shear properties of the materials. This is particularly relevant for oil spill scenarios, where environmental conditions can fluctuate, indicating the necessity to design sorbents that maintain their integrity under diverse conditions.

Additionally, Bailly et al. [78] focused on the strain rate-dependent behavior of vinyl nitrile foam under compression and shear, providing insights into how these materials respond to dynamic loading conditions, essential for simulating real-world oil spill scenarios where foam sorbents may encounter rapid changes in stress due to waves or mechanical recovery efforts. The research conducted by Ling et al. [79] on expanded polystyrene (EPS) foam under combined compression-shear loading demonstrated that the mechanical behavior of foams can vary significantly depending on the type of loading applied. The study's results indicate that foams must be engineered to withstand complex loading scenarios typical in oil spill recovery operations.

These investigations collectively highlight the critical need for understanding the mechanical stresses that bio-based foam sorbents encounter during oil spill recovery. The insights gained can guide manufacturers in designing sorbents that are robust and durable, capable of withstanding compression, tension, and shear stresses, thereby enhancing their performance and longevity, ensuring effective oil recovery in diverse conditions.

3.4.2 Strength and Durability

Mechanical strength is essential for sorbents that are repeatedly used and must endure rough handling during oil spill cleanups. Compressive strength is vital for resisting crushing under weight or other forces, while tensile strength is important for preventing the sorbent from tearing when stretched. The following discussion explores how advanced material formulations improve mechanical strength and how this relates to the durability and reusability of sorbents in harsh marine environments, emphasizing the materials' fiber bonding and matrix integrity, which contribute to their overall structural robustness.

Recent studies have illuminated the importance of enhancing fiber interfacial adhesion in composite foams to increase both compressive and tensile strengths. For instance, Viju et al. [80] examined the oil sorption properties of cotton comber noil blended with recycled polyester in needle-punched nonwoven fabrics, finding that bonding between fibers significantly influences the sorbent's mechanical properties. This improvement in fiber interfacial adhesion can enhance the sorbent's tensile and compressive strengths, crucial for effective oil spill recovery.

Similarly, Sahu et al. [81] and Ghadei et al. [82] have demonstrated the critical roles of structural integrity and advanced material formulations. Sahu's study on used cooking oil-derived porous polysulfide materials showed how the foam's structural integrity is essential during oil absorption and recovery processes, with advanced materials improving fiber-matrix interactions. Ghadei's research on dandelion-like structured foam highlighted how unique structural designs contribute to improved compressive strength, underscoring the multifunctional potential of these materials alongside their oil/water separation capabilities.

Further reinforcing the significance of material modifications, Zimmermann et al. [83] revealed that acetylated microfibrillated cellulose enhances the interfacial adhesion within the foam matrix, increasing durability and mechanical strength. Mukherjee et al. [84] also noted the beneficial impact of nanofillers on bio-based foams, enhancing mechanical properties and oil adsorption capacity suitable for repeated use in spill scenarios.

Additionally, Topuz et al. [85] and Visco et al. [86] explored the production of advanced materials like hierarchically porous electrospun nanofibrous mats and polyurethane foams loaded with carbon nanofibers. These studies demonstrated that strong interactions between fibers and the polymer matrix, as well as the incorporation of nanofillers, significantly improve the mechanical properties of foam sorbents, contributing to their mechanical durability under fatigue conditions and enhancing their compressive and tensile strengths.

Together, these studies illustrate that advancements in fiber bonding and matrix integrity are crucial for enhancing the mechanical strength of bio-based foam sorbents used in oil spill cleanups. The incorporation of advanced materials and chemical modifications can significantly improve the compressive and tensile strengths of these sorbents, ensuring their durability and reusability in harsh marine environments. Continued research in this area will be essential for developing effective and sustainable solutions for oil spill remediation.

3.4.3 Elasticity and Flexibility

The elasticity and flexibility of sorbents play a significant role in their deployment capabilities. Flexible materials can adapt to irregular surfaces and maintain contact with the spilled oil, which is crucial for efficient absorption. Elasticity refers to the ability of the sorbent to return to its original shape after deformation, which is important for operational handling and reusability. This discussion will delve into the material science behind designing sorbents that balance flexibility with sufficient mechanical strength to withstand the rigors of deployment. Innovations in polymer blends and foam structures that provide enhanced elasticity without compromising the sorbent's structural integrity will be highlighted.

The role of elasticity and flexibility in the performance of bio-based foam sorbents for oil spill recovery is critical, particularly in how these materials adapt to irregular surfaces and maintain contact with spilled oil. Piperopoulos et al. [87] investigated the sorption kinetics of carbon nanotube-based composite foams, finding that the interconnected bubble structure within the foam significantly enhances sorption performance, particularly in terms of flexibility and adaptability to irregular surfaces. High-porosity foams, which are typically low in density, allow for faster oil absorption, demonstrating that flexibility is essential for effective oil recovery.

In studies discussed in previous sections, such as those by Calabrese et al. [88] and Tomon et al. [20], the focus on advanced material formulations highlights significant innovations in sorbent technologies. Each of these studies contributes to our understanding of how sorbents can be engineered to optimize performance in oil spill recovery scenarios. Calabrese et al. explored polydimethylsiloxane (PDMS) foams filled with thermally treated bentonite, highlighting their ability to float on spilled oil while maintaining structural integrity. The squeezing capacity of these foams allows them to recover their shape after deformation, which is crucial for operational handling and reusability. Tomon on coconut oil-based foams demonstrated that inherent hydrophobic properties contribute to the flexibility and oil sorption capabilities of the material. The study noted that the viscosity of the oil influences the foam's absorption capacity, with flexible materials being better suited to adapt to the varying viscosities of different oils. This adaptability is crucial for maintaining contact with spilled oil and ensuring efficient absorption.

Additionally, Wang et al. [89] developed polypropylene foams with advanced solvent-evaporation-induced structures that exhibit superhydrophobic properties, allowing them to selectively adsorb oil underwater without absorbing water. Their findings indicated that after multiple cycles of centrifugation and adsorption, the foams maintained over 90% of their original oil absorption capacity. This resilience is attributed to the elasticity of the foam, which enables it to return to its original shape after deformation, enhancing its reusability in oil spill scenarios.

Cifarelli et al. [90] and Perera et al. [52] also made significant contributions to this field. Cifarelli et al. focused on flexible polyurethane foams from epoxidized vegetable oils, showing that larger pore sizes contribute to better elastic properties, allowing the foams to recover their original shape after compression. Perera et al. investigated castor oil-based polyurethane foams grafted with modified diatomite, emphasizing the importance of flexibility and low density in enhancing the sorbent's performance. The flexibility of these foams allows them to conform to various surface geometries, improving their effectiveness in capturing oil spills.

These studies collectively underscore the importance of designing sorbents that are not only flexible and elastic but also robust enough to withstand the conditions encountered during oil spill recoveries. The insights from these investigations guide manufacturers in developing sorbents that offer both operational efficiency and sustainability in harsh marine environments.

3.4.4 Impact of Environmental Factors

The mechanical properties of bio-based foam sorbents used in oil spill recovery are significantly influenced by environmental factors such as temperature, salinity, and water chemistry. Recent studies have explored how these conditions affect the performance of these materials, particularly in terms of brittleness, corrosion, and swelling, as well as strategies to enhance their resilience.

Low temperatures, for example, can increase brittleness in biobased foam materials, thereby reducing their mechanical strength. Wang et al. [91] highlight that the mechanical properties of polyurethane foams are particularly sensitive to temperature variations, with colder conditions typically resulting in a more brittle material structure. This brittleness can compromise the foam's ability to absorb oil effectively during spill recovery operations. D'Arienzo et al. [92] discuss how environmental aging, including temperature fluctuations, affects the deformation mechanisms of flexible polyurethane foams, indicating that mechanical performance can degrade under cold conditions. Abd Rahim et al. [93] also emphasize that foam performance in enhanced oil recovery is heavily influenced by temperature, suggesting that maintaining optimal thermal conditions is essential for foam stability and effectiveness.

In terms of salinity, high levels can adversely affect the structural integrity of foam sorbents. Obisesan et al. [94] investigate the impact of salt on the stability of aqueous foams, revealing that increased salinity can lead to enhanced drainage and destabilization of foam structures. This finding is particularly relevant for biobased foams, which may be more susceptible to ionic interactions that can cause swelling or corrosion. Mi et al. [95] note that certain polymer foams exhibit resistance to corrosive liquids, which is critical for maintaining functionality in saline environments.

The development of biobased foam materials with enhanced mechanical properties and corrosion resistance is an area of active research. Liang et al. [96] discuss how external factors, including salinity and temperature, can significantly influence foam properties, suggesting that optimizing these parameters is essential for improving foam resilience. Fang et al. [97] explore the use of modified polyols to create high-performance biodegradable foams that maintain structural integrity under various environmental conditions. Calabrese et al. [88] highlight the benefits of using polydimethylsiloxane (PDMS)-based foams, which exhibit excellent chemical stability and hydrophobicity, making them suitable for oil spill remediation.

These insights illustrate the critical need to design sorbents with materials that can withstand environmental challenges. Enhancing the resilience of sorbents through corrosion-resistant materials and treatments ensures their functionality remains consistent in a variety of spill conditions. Future research will continue to focus on optimizing foam formulations and incorporating treatments that enhance performance under adverse conditions.

3.5 Thermal Stability

Thermal stability is a key determinant of bio-based foam sorbents' effectiveness in oil spill cleanup operations. The capability of a sorbent to preserve its structure and functionality under varying temperatures directly impacts its ability to absorb and retain oil. When exposed to high temperatures, which may occur during oil recovery or in certain environmental conditions, the thermal stability of the foam sorbent becomes crucial. Materials that can withstand elevated temperatures without structural or functional degradation are more efficient in oil spill cleanup due to their sustained absorption capacity and durability.

3.5.1 Structural Integrity Under Thermal Stress

Foam sorbents, particularly those with bio-based compositions like polyurethane derived from palm oil, exhibit high thermal stability, maintaining their structure and oil absorption properties even when exposed to heat. For example, polyurethane foam synthesized from palm oil derivatives has been shown to retain its integrity at elevated temperatures, thus ensuring effective performance in oil recovery processes [98]. This stability is largely due to its chemical structure, where polyols from palm oil react with isocyanates to form urethane linkages. These linkages provide the foam with strength and resistance to thermal degradation [34]. Additionally, the presence of saturated fatty acids from palm oil enhances thermal stability, allowing these foams to perform consistently without significant performance loss [41].

3.5.2 Enhancement Through Additives

The incorporation of nanoparticles or fillers is a strategy to improve the thermal stability of foam sorbents. For instance, TiO_2 nanoparticles have been shown to enhance the thermal properties of polyurethane foams, enabling them to retain functionality during prolonged exposure to heat [99]. Such enhancements are valuable in applications where foam sorbents may encounter variable temperatures, such as in oil spill cleanups in warmer climates or industrial settings. Foams with high porosity and well-defined open-cell structures dissipate heat more efficiently, mitigating localized overheating and preserving structural integrity, which is essential for buoyancy and oil absorption efficiency [100].

3.5.3 Influence on Buoyancy and Oil Sorption

Thermal stability also affects the buoyancy of foam sorbents, a crucial feature for performance in aquatic environments. Sorbents that lose their structural integrity due to thermal degradation may become waterlogged, reducing their ability to float and absorb oil from the surface. The preservation of buoyancy is particularly important when sorbents need to remain at the water's surface to maximize oil uptake [34]. Therefore, foam sorbents that demonstrate high thermal stability are more effective in oil spill applications due to their consistent buoyancy and functional durability.

3.5.4 Impact on Reusability and Sustainability

Thermal stability further contributes to the reusability of foam sorbents. Materials capable of withstanding multiple heating and cooling cycles without performance degradation are more sustainable and cost-effective, particularly in large-scale cleanup operations. Sorbents that endure thermal regeneration processes to rejuvenate oil absorption capacity allow for repeated use, reducing waste and enhancing remediation efficiency [4]. Ultimately, thermal stability is a pivotal attribute for foam sorbents, influencing not only their immediate oil absorption efficiency but also their longevity, environmental resilience, and contribution to sustainable oil spill cleanup efforts [101].

3.6 Environmental Conditions

Environmental conditions greatly impact the effectiveness of foam sorbents in oil spill cleanup, with factors such as temperature, salinity, and pH influencing sorbent structure, oil absorption capacity, and overall performance. Understanding how these conditions interact with foam sorbents is essential to optimizing their design and application across various environmental contexts.

3.6.1 Temperature Effects

Temperature is a key environmental factor affecting foam sorbent performance. Elevated temperatures can increase the fluidity of oil, making it easier for sorbents to absorb. However, high temperatures may also cause thermal degradation in the foam itself, diminishing its mechanical strength and oil retention capacity [102]. For instance, polyurethane foams synthesized from bio-based polyols, such as those derived from palm oil, show varying levels of thermal stability. While effective at moderate temperatures, prolonged exposure to high temperatures may weaken the structural integrity of these foams, thereby reducing their performance in oil absorption [102].

3.6.2 Salinity Impact

Salinity, especially in marine environments, can significantly influence foam sorbent efficiency. The presence of saltwater can alter surface tension and the wettability of the sorbent, potentially decreasing its oil absorption capacity. Research suggests that certain bio-based foams, such as those made from coconut oil, may have reduced efficiency in saline conditions due to altered hydrophobic properties [103]. This reduction is attributed to interactions between salt ions and the foam's surface, which disrupt the oil-sorbent interface and lower absorption efficiency.

3.6.3 pH Variations

The pH of the surrounding environment also affects foam sorbent performance, as acidic or alkaline conditions can modify the surface chemistry of the foam, impacting its hydrophobic and oleophilic properties. For instance, a highly alkaline environment may cause hydrolysis in some foam materials, lowering their oil absorption efficacy. To counter these pH effects, researchers have developed surface modifications, such as hydrophobic coatings or additives, to improve foam sorbent stability across a range of pH levels, helping maintain effectiveness in diverse conditions [104].

3.6.4 Strategies to Enhance Environmental Resilience

To mitigate the adverse effects of environmental conditions, several enhancement strategies can be applied to foam sorbents. One approach involves incorporating additives that improve resilience. For example, adding nanoparticles like titanium dioxide (TiO_2) has been shown to enhance the thermal stability and oil absorption capacity of polyurethane foams, allowing them to maintain performance under fluctuating conditions [42]. These modifications help ensure that the foam sorbents remain effective in oil spill cleanup across varying environments.

Another strategy is the development of hybrid or composite materials that blend the advantages of different sorbents. Combining bio-based foams with inorganic materials like bentonite can enhance structural integrity and oil retention, making these materials more robust in challenging environmental settings [88]. This approach not only enhances performance but also aligns with sustainability by incorporating renewable resources.

Environmental factors such as temperature, salinity, and pH play a crucial role in determining the performance of foam sorbents in oil spill remediation. By employing strategies like additive incorporation and hybrid material development, the resilience and effectiveness of bio-based foam sorbents can be significantly improved. These enhancements enable foam sorbents to operate successfully in diverse environmental scenarios, ensuring more efficient and sustainable oil spill cleanup.

3.7 Operational Parameters

Operational parameters are critical in determining the effectiveness of foam sorbents in oil spill cleanup. These factors, including contact time with oil, oil viscosity, environmental temperature, and water flow dynamics, directly impact the sorbent's ability to absorb oil and the rate at which it can do so. By understanding these parameters, sorbent design and deployment can be optimized for better performance.

3.7.1 *Contact Time with Oil*

Contact time is one of the most important operational parameters for oil absorption. Extended contact allows more interaction between the sorbent and oil molecules, generally enhancing absorption capacity. However, in dynamic environments like moving or wave-affected water, maintaining effective contact can be challenging if the sorbent is not securely positioned, potentially reducing its performance. For example, bio-based foam sorbents from coconut oil demonstrate high oil absorption capacity, but their effectiveness can diminish in turbulent water if they are not given sufficient time to interact with the oil [103].

3.7.2 *Oil Viscosity*

Oil viscosity significantly affects sorbent performance. High-viscosity oils, such as heavy crude oil, present absorption challenges as they spread slowly over the sorbent surface, resulting in lower absorption rates and efficiency [105]. To address this, researchers have enhanced foam sorbents with additives to improve oleophilicity, making them more effective for viscous oils. For instance, applying superhydrophobic coatings to polyurethane foams can increase their capacity to absorb high-viscosity oils, even under challenging conditions [48].

3.7.3 *Temperature Influence*

Temperature plays a dual role in sorbent performance. Elevated temperatures increase oil fluidity, which can facilitate absorption. However, excessive heat risks degrading the foam material, potentially reducing effectiveness [2]. A balanced approach is required to capitalize on increased oil fluidity while minimizing the risk of thermal degradation. Bio-based foam sorbents modified with flame retardants or stabilizing agents offer improved thermal stability, enhancing performance at higher temperatures without compromising material integrity [106].

3.7.4 *Water Flow Dynamics*

Flow dynamics in water, particularly in turbulent environments, can impact foam sorbent performance by displacing the material and reducing oil contact. In these conditions, anchored or weighted foam sorbents may offer greater stability, ensuring they remain in place despite water movement [107]. Additionally, designing foam sorbents with enhanced buoyancy and stability can help them maintain consistent contact with oil in dynamic environments, supporting effective cleanup.

Operational parameters such as contact time, oil viscosity, temperature, and water flow dynamics are crucial to foam sorbent performance in oil spill remediation. By developing bio-based foam sorbents that can adapt to these conditions, their effectiveness can be maximized, resulting in more efficient and reliable oil spill cleanup.

3.8 Recyclability and Reusability

Recyclability and reusability are essential factors that improve the performance and sustainability of foam sorbents in oil spill cleanup. These characteristics allow foam sorbents to be used multiple times, reducing waste and enhancing the economic feasibility of large-scale cleanup applications. The ability to recycle and reuse foam sorbents lowers the demand for frequent replacement and aligns with sustainability goals.

3.8.1 Desorption Efficiency

Desorption efficiency, or a sorbent's ability to release absorbed oil after saturation, is crucial for its reusability. High desorption efficiency enables the recovery of absorbed oil, allowing the foam to be reused without significant loss in performance. This characteristic is especially valuable in environmental remediation, where the focus is not only on oil removal but also on minimizing the environmental footprint of the cleanup materials themselves [34].

3.8.2 Durability

Durability refers to the sorbent's capacity to withstand repeated absorption and desorption cycles without degrading. Durable foam sorbents maintain structural integrity and performance over time, which is key for effective reuse [108]. Expanded polystyrene (EPS) is an example of a sorbent with strong recyclability and reusability due to its robust structure and resistance to degradation [99]. This resilience allows EPS foams to be reused in oil spill cleanup operations, enhancing their overall effectiveness.

3.8.3 Bio-based Polyurethane Foam Example

Polyurethane foam derived from renewable resources, such as palm oil, is an example of a bio-based sorbent with excellent recyclability and reusability. Engineered for high oil absorption capacity and durability, these foams are designed to retain oil effectively while maintaining good mechanical properties [107]. Their chemical composition also enables efficient desorption of absorbed oil, supporting multiple reuse cycles. The polymer network in palm oil-based polyurethane foams, along with their thermal stability, allows them to withstand cleaning processes like squeezing or thermal regeneration, restoring the foam's functionality after each cycle [109]. Studies indicate that these foams retain high performance even after multiple cycles, often achieving desorption efficiencies above 75%, making them both economically viable and sustainable for large-scale use [110].

3.8.4 Enhancement Strategies

To further improve recyclability and reusability, researchers are exploring strategies that enhance the mechanical properties and thermal stability of foam sorbents. One approach involves incorporating additives, such as nanoparticles, that reinforce the foam's durability, allowing it to withstand more cycles without degradation [111]. Additionally, surface modifications can improve the foam's hydrophobic and oleophilic properties, boosting both oil absorption and desorption efficiency. Coatings that increase the foam's contact angle with oil enhance its oil absorption while repelling water. For instance, organosilane coatings have significantly improved the oil absorption capacity of polyurethane foams, making them more effective in oil spill cleanup [98]. These modifications not only improve initial absorption rates but also facilitate easier oil desorption during recycling, enhancing the foam's overall reusability.

The recyclability and reusability of foam sorbents are vital for efficient and sustainable oil spill remediation. By focusing on high desorption efficiency, durability, and innovative enhancements, bio-based foam sorbents can be optimized for repeated use, reducing waste and supporting large-scale environmental cleanup efforts.

3.9 Environmental Impact

The environmental impact of foam sorbents is a key consideration in their use for oil spill remediation. Designed to absorb and retain spilled oil, foam sorbents are often exposed to harsh conditions, including variable temperatures, water exposure, and extended oil contact, all of which can influence their efficiency. Evaluating their environmental impact involves not only assessing their oil absorption capacity but also examining their biodegradability, toxicity, and overall ecological footprint post-use.

3.9.1 Biodegradability

Biodegradability is a critical factor in evaluating the environmental impact of foam sorbents. Biodegradable materials break down naturally, reducing the long-term ecological footprint of sorbents after they have been deployed. Bio-based foam sorbents derived from natural sources, such as coconut oil, sunflower oil, palm olein, lignin, cellulose, and polylactic acid (PLA), generally degrade more readily than petroleum-based foams. For example, polyurethane foams made from coconut and sunflower oil demonstrate impressive oil absorption while breaking down into non-toxic components, thereby reducing waste accumulation in the environment [18]. By ensuring these materials do not persist post-cleanup, biodegradability prevents additional environmental pollution.

3.9.2 Toxicity

Toxicity is another vital aspect of a foam sorbent's environmental impact. To be environmentally safe, a sorbent must be non-toxic to both aquatic and terrestrial ecosystems, particularly in oil spill scenarios. Many synthetic foams release harmful chemicals as they degrade, posing risks to marine life and local ecosystems. In contrast, bio-based sorbents, such as those made from lignin or cellulose, typically exhibit low toxicity due to their plant-based origins, which do not release harmful substances upon degradation [24]. For instance, lignin-based foams, valued for their oil absorption properties, are shown to be minimally toxic to marine organisms, supporting both ecological safety and effective oil spill remediation [112].

3.9.3 Effectiveness and Material Properties

The effectiveness of bio-based foam sorbents in cleanup operations is influenced by their chemical structure and their hydrophobic and oleophilic characteristics. Materials like PLA and rubber-based foams repel water while attracting oil, making them ideal for oil absorption. Their hydrophobicity ensures buoyancy, a critical property for maintaining contact with oil slicks on the water surface. For instance, PLA-based foam sorbents combine high oil sorption efficiency with biodegradability, providing a low-toxicity, environmentally friendly solution for oil spill remediation [66].

3.9.4 Enhancements with Sustainable Additives

Incorporating sustainable additives can further improve the environmental impact of foam sorbents by enhancing biodegradability and reducing their ecological footprint. Additives such as natural fibers, chitosan, or nanocellulose can improve foam properties like porosity and strength, while also supporting faster decomposition after use. These improvements collectively enhance the sustainability of foam sorbents, allowing them to perform effectively in oil cleanup with minimal long-term environmental impact [19].

Polyurethane foam derived from palm olein is a notable example of an environmentally friendly sorbent. This foam not only demonstrates excellent thermal stability, buoyancy, and oil absorption but is also biodegradable. The fatty acid composition in palm olein contributes to a stable foam structure while promoting biodegradation upon disposal, making it an ideal choice for eco-conscious oil spill cleanup [113].

Bio-based foam sorbents made from materials such as coconut oil, sunflower oil, lignin, cellulose, PLA, and rubber show promise in oil spill remediation while minimizing environmental impact. Their biodegradability supports natural decomposition, reducing waste and ecological damage, while low toxicity safeguards aquatic and terrestrial life. Combined with effective oil absorption, buoyancy, and enhancements from sustainable additives, bio-based foam sorbents offer a sustainable solution for efficient and environmentally safe oil spill remediation.

References

1. Doshi, B., Sillanpää, M., Kalliola, S.: A review of bio-based materials for oil spill treatment. Water Res. **135**, 262–277 (2018). https://doi.org/10.1016/j.watres.2018.02.034
2. Kathir, K., Carroll, M.: From trash to treasure: a sustainable approach to oil spill clean-up. J. Emerg. Investig. (2023). https://doi.org/10.59720/21-018
3. Cywar, R.M., Rorrer, N.A., Hoyt, C.B., Beckham, G.T., Chen, E.Y.-X.: Bio-based polymers with performance-advantaged properties. Nat. Rev. Mater. **7**(2), 83–103 (2021). https://doi.org/10.1038/s41578-021-00363-3
4. Ge, J., Zhao, H., Zhu, H., Huang, J., Shi, L., Yu, S.: Advanced sorbents for oil-spill cleanup: recent advances and future perspectives. Adv. Mater. **28**(47), 10459–10490 (2016). https://doi.org/10.1002/adma.201601812
5. Ge, J., et al.: Pumping through porous hydrophobic/oleophilic materials: an alternative technology for oil spill remediation. Angew. Chemie **126**(14), 3686–3690 (2014). https://doi.org/10.1002/ange.201310151
6. Yazdani, M., Virolainen, E., Conley, K., Vahala, R.: Chitosan–Zinc(II) complexes as a bio-sorbent for the adsorptive abatement of phosphate: mechanism of complexation and assessment of adsorption performance. Polymers (Basel) **10**(1), 25 (2017). https://doi.org/10.3390/polym10010025
7. Ifelebuegu, A.O., Johnson, A.: Nonconventional low-cost cellulose- and keratin-based biopolymeric sorbents for oil/water separation and spill cleanup: a review. Crit. Rev. Environ. Sci. Technol. **47**(11), 964–1001 (2017). https://doi.org/10.1080/10643389.2017.1318620

8. Kelle, H.I., Eboatu, A.N.: Determination of the viability of chicken feather as oil spill clean-up sorbent for crude oil and its lower fractions. J. Appl. Sci. Environ. Manag. **22**(2), 267 (2018). https://doi.org/10.4314/jasem.v22i2.19
9. Du, W., Han, X., Li, Z., Li, Y., Li, L., Wang, K.: Oil sorption behaviors of porous polydimethylsiloxane modified collagen fiber matrix. J. Appl. Polym. Sci. **132**(44), (2015). https://doi.org/10.1002/app.42727
10. Singh, V., Jinka, S., Hake, K., Parameswaran, S., Kendall, R.J., Ramkumar, S.: Novel natural sorbent for oil spill cleanup. Ind. Eng. Chem. Res. **53**(30), 11954–11961 (2014). https://doi.org/10.1021/ie5019436
11. Alinejad, M., et al.: Lignin-based polyurethanes: opportunities for bio-based foams, elastomers, coatings and adhesives. Polymers (Basel) **11**(7), 1202 (2019). https://doi.org/10.3390/polym11071202
12. Luo, S., Gao, L., Guo, W.: Effect of incorporation of lignin as bio-polyol on the performance of rigid lightweight wood–polyurethane composite foams. J. Wood Sci. **66**(1), 23 (2020). https://doi.org/10.1186/s10086-020-01872-5
13. Yang, J., Ching, Y., Chuah, C.: Applications of lignocellulosic fibers and lignin in bioplastics: a review. Polymers (Basel) **11**(5), 751 (2019). https://doi.org/10.3390/polym11050751
14. Quan, P., Kiziltas, A., Gondaliya, A., Siahkamari, M., Nejad, M., Xie, X.: Kraft lignin with improved homogeneity recovered directly from black liquor and its application in flexible polyurethane foams. ACS Omega **7**(19), 16705–16715 (2022). https://doi.org/10.1021/acsomega.2c01206
15. Korányi, T.I., Fridrich, B., Pineda, A., Barta, K.: Development of 'lignin-first' approaches for the valorization of lignocellulosic biomass. Molecules **25**(12), 2815 (2020). https://doi.org/10.3390/molecules25122815
16. Henry, C., Nejad, M.: Lignin-based low-density rigid polyurethane/polyisocyanurate foams. Ind. Eng. Chem. Res. **62**(18), 6865–6873 (2023). https://doi.org/10.1021/acs.iecr.2c03728
17. Gama, N.V., Ferreira, A., Barros-Timmons, A.: Polyurethane foams: past, present, and future. Materials (Basel) **11**(10), 1841 (2018). https://doi.org/10.3390/ma11101841
18. Li, H., Kognou, A.L.M., Jiang, Z., Qin, W., Xu, C.: Production of bio-polyurethane (BPU) foams from greenhouse/agricultural wastes, and their biodegradability. Biofuels Bioprod. Biorefin. **16**(3), 826–837 (2022). https://doi.org/10.1002/bbb.2344
19. Purwanto, N.S., Chen, Y., Torkelson, J.M.: Reprocessable, bio-based, self-blowing non-isocyanate polyurethane network foams from cashew nutshell liquid. ACS Appl. Polym. Mater. **5**(8), 6651–6661 (2023). https://doi.org/10.1021/acsapm.3c01196
20. Tomon, T.R.B., et al.: A novel naturally superoleophilic coconut oil-based foam with inherent hydrophobic properties for oil and grease sorption. Sci. Rep. **14**(1), 14223 (2024). https://doi.org/10.1038/s41598-024-64178-2
21. Kurańska, M., Leszczyńska, M., Malewska, E., Prociak, A., Ryszkowska, J.: Implementation of circular economy principles in the synthesis of polyurethane foams. Polymers (Basel) **12**(9), 2068 (2020). https://doi.org/10.3390/polym12092068
22. Wang, Y., et al.: Hybrid electrospun porous fibers of poly(lactic acid) and nano ZIF-8@C600 as effective degradable oil sorbents. J. Chem. Technol. Biotechnol. **95**(3), 730–738 (2020). https://doi.org/10.1002/jctb.6256
23. Alassod, A., Gibril, M., Islam, S.R., Huang, W., Xu, G.: Polypropylene/lignin blend monoliths used as sorbent in oil spill cleanup. Heliyon **6**(9), e04591 (2020). https://doi.org/10.1016/j.heliyon.2020.e04591
24. Bhagwat, S.B., Jaspal, D.: Application of biodegradable materials in oil-spill cleanup: recent advances and future perspectives. Adv. Eng. Mater. **25**(20) (2023). https://doi.org/10.1002/adem.202300315
25. Zeng, M., et al.: Highly biocompatible, underwater superhydrophilic and multifunctional biopolymer membrane for efficient oil-water separation and aqueous pollutant removal. ACS Sustain. Chem. Eng. **6**(3), 3879–3887 (2018). https://doi.org/10.1021/acssuschemeng.7b04219

26. Sharma, M., Tellili, N., Kacem, I., Rouissi, T.: Microbial biopolymers: from production to environmental applications—a review. Appl. Sci. **14**(12), 5081 (2024). https://doi.org/10.3390/app14125081
27. Cao, S., Bate, B., Hu, J., Jung, J.: Engineering behavior and characteristics of water-soluble polymers: implication on soil remediation and enhanced oil recovery. Sustainability **8**(3), 205 (2016). https://doi.org/10.3390/su8030205
28. Balíková, K., Farkas, B., Matúš, P., Urík, M.: Prospects of biogenic xanthan and Gellan in removal of heavy metals from contaminated waters. Polymers (Basel) **14**(23), 5326 (2022). https://doi.org/10.3390/polym14235326
29. Jung, J., Hu, J.: Characterization of polyethylene oxide and sodium alginate for oil contaminated-sand remediation. Sustainability **9**(1), 62 (2017). https://doi.org/10.3390/su9010062
30. Wu, F., Pickett, K., Panchal, A., Liu, M., Lvov, Y.: Superhydrophobic polyurethane foam coated with polysiloxane-modified clay nanotubes for efficient and recyclable oil absorption. ACS Appl. Mater. Interfaces **11**(28), 25445–25456 (2019). https://doi.org/10.1021/acsami.9b08023
31. Hoang, P.H., Dat, H.T., Cuong, T.D., Dien, L.Q.: Pretreatment of coir lignocellulose for preparation of a porous coir–polyurethane composite with high oil adsorption capacity. RSC Adv. **12**(24), 14976–14985 (2022). https://doi.org/10.1039/D2RA01349E
32. Than, K.T., Yu, L.J., Chen, R.S., Tarawneh, M.A., Lai, N.Y.G.: Density, water, and swelling characteristics of cellular palm kernel oil bio-polyol modified polyurethane toward more sustainable growing media application. J. Appl. Polym. Sci. **139**(37) (2022). https://doi.org/10.1002/app.52879
33. Song, W., Barber, K., Lee, K.-Y.: Heat-induced bubble expansion as a route to increase the porosity of foam-templated bio-based macroporous polymers. Polymer (Guildf) **118**, 97–106 (2017). https://doi.org/10.1016/j.polymer.2017.04.058
34. Zhu, H., Qiu, S., Jiang, W., Wu, D., Zhang, C.: Evaluation of electrospun polyvinyl chloride/polystyrene fibers as sorbent materials for oil spill cleanup. Environ. Sci. Technol. **45**(10), 4527–4531 (2011). https://doi.org/10.1021/es2002343
35. Xiao, W., et al.: Fabrication of foam-like oil sorbent from polylactic acid and calotropis gigantea fiber for effective oil absorption. J. Clean. Prod. **278**, 123507 (2021). https://doi.org/10.1016/j.jclepro.2020.123507
36. Mardiyati, Y., Fauza, A.N., Rachman, O.A., Steven, S., Santosa, S.P.: A silica-lignin hybrid filler in a natural rubber foam composite as a green oil spill absorbent. Polymers (Basel) **14**(14), 2930 (2022). https://doi.org/10.3390/polym14142930
37. Liu, H., Geng, B., Chen, Y., Wang, H.: Review on the aerogel-type oil sorbents derived from nanocellulose. ACS Sustain. Chem. Eng. **5**(1), 49–66 (2017). https://doi.org/10.1021/acssuschemeng.6b02301
38. Lorevice, M.V., Mendonça, E.O., Orra, N.M., Borges, A.C., Gouveia, R.F.: Porous cellulose nanofibril-natural rubber latex composite foams for oil and organic solvent absorption. ACS Appl. Nano Mater. **3**(11), 10954–10965 (2020). https://doi.org/10.1021/acsanm.0c02203
39. Kurańska, M., Barczewski, R., Barczewski, M., Prociak, A., Polaczek, K.: Thermal insulation and sound absorption properties of open-cell polyurethane foams modified with bio-polyol based on used cooking oil. Materials (Basel) **13**(24), 5673 (2020). https://doi.org/10.3390/ma13245673
40. Gu, R., Konar, S., Sain, M.: Preparation and characterization of sustainable polyurethane foams from soybean oils. J. Am. Oil Chem. Soc. **89**(11), 2103–2111 (2012). https://doi.org/10.1007/s11746-012-2109-8
41. Zimmermann, M.V.G., Zattera, A.J., Fenner, B.R., Santana, R.M.C.: Sorbent system based on organosilane-coated polyurethane foam for oil spill clean up. Polym. Bull. **78**(3), 1423–1440 (2021). https://doi.org/10.1007/s00289-020-03169-5
42. Wei, Q., Oribayo, O., Feng, X., Rempel, G.L., Pan, Q.: Synthesis of polyurethane foams loaded with TiO_2 nanoparticles and their modification for enhanced performance in oil spill cleanup. Ind. Eng. Chem. Res. **57**(27), 8918–8926 (2018). https://doi.org/10.1021/acs.iecr.8b01037

43. Wang, Z., Barford, J.P., Hui, C.W., McKay, G.: Kinetic and equilibrium studies of hydrophilic and hydrophobic rice husk cellulosic fibers used as oil spill sorbents. Chem. Eng. J. **281**, 961–969 (2015). https://doi.org/10.1016/j.cej.2015.07.002
44. Zamparas, M., Tzivras, D., Dracopoulos, V., Ioannides, T.: Application of sorbents for oil spill cleanup focusing on natural-based modified materials: a review. Molecules **25**(19), 4522 (2020). https://doi.org/10.3390/molecules25194522
45. Khalid, F.E., et al.: Guinea grass (*Megathyrsus maximus*) fibres as sorbent in diesel bioremediation. Sustainability **15**(5), 3904 (2023). https://doi.org/10.3390/su15053904
46. Liu, X., Meng, R., Xing, Q., Lou, M., Chao, H., Bing, L.: Assessing oil spill risk in the Chinese Bohai Sea: a case study for both ship and platform related oil spills. Ocean Coast. Manag. **108**, 140–146 (2015). https://doi.org/10.1016/j.ocecoaman.2014.08.016
47. Vasilyeva, J., Vasekha, M., Tyulyaev, V.: Evaluation of the efficiency of sorbents for accidental oil spill response in the Arctic waters. J. Min. Inst., vol. Online fir (2023). https://doi.org/10.31897/PMI.2023.14
48. Kim, H., Zhang, G., Chung, T.C.M., Nam, C.: A role for newly developed sorbents in remediating large-scale oil spills: reviewing recent advances and beyond. Adv. Sustain. Syst. **6**(1) (2022). https://doi.org/10.1002/adsu.202100211
49. Yang, W.-J., et al.: Recent progress in bio-based aerogel absorbents for oil/water separation. Cellulose **26**(11), 6449–6476 (2019). https://doi.org/10.1007/s10570-019-02559-x
50. Navarathna, C.M., et al.: Biochar adsorbents with enhanced hydrophobicity for oil spill removal. ACS Appl. Mater. Interfaces **12**(8), 9248–9260 (2020). https://doi.org/10.1021/acsami.9b20924
51. Nam, C., et al.: Practical oil spill recovery by a combination of polyolefin absorbent and mechanical skimmer. ACS Sustain. Chem. Eng. **6**(9), 12036–12045 (2018). https://doi.org/10.1021/acssuschemeng.8b02322
52. Perera, H.J., Goyal, A., Alhassan, S.M., Banu, H.: Biobased castor oil-based polyurethane foams grafted with octadecylsilane-modified diatomite for use as eco-friendly and low-cost sorbents for crude oil clean-up applications. Polymers (Basel) **14**(23), 5310 (2022). https://doi.org/10.3390/polym14235310
53. Malewska, E., Bąk, S., Prociak, A.: Effect of different concentration of rapeseed-oil-based polyol and water on structure and mechanical properties of flexible polyurethane foams. J. Appl. Polym. Sci. **132**(35) (2015). https://doi.org/10.1002/app.42372
54. Coste, G., Negrell, C., Averous, L., Caillol, S.: Green synthesis of biobased soft foams by the Aza-Michael reaction. ACS Sustain. Chem. Eng. **10**(26), 8549–8558 (2022). https://doi.org/10.1021/acssuschemeng.2c01885
55. Furtwengler, P., Matadi Boumbimba, R., Sarbu, A., Avérous, L.: Novel rigid polyisocyanurate foams from synthesized biobased polyester polyol with enhanced properties. ACS Sustain. Chem. Eng. **6**(5), 6577–6589 (2018). https://doi.org/10.1021/acssuschemeng.8b00380
56. Zhang, J., Cheng, Y., Xu, C., Gao, M., Zhu, M., Jiang, L.: Hierarchical interface engineering for advanced nanocellulosic hybrid aerogels with high compressibility and multifunctionality. Adv. Funct. Mater. **31**(19) (2021). https://doi.org/10.1002/adfm.202009349
57. Bidgoli, H., Mortazavi, Y., Khodadadi, A.A.: A functionalized nano-structured cellulosic sorbent aerogel for oil spill cleanup: synthesis and characterization. J. Hazard. Mater. **366**, 229–239 (2019). https://doi.org/10.1016/j.jhazmat.2018.11.084
58. Soares, L.F., César dos Santos, J., Araújo de Freitas, V.A., Dutra Pereira, R.B., Panzera, T.H., Scarpa, F.: Castor-oil biobased foam: the effect of the composition on the physical and mechanical properties via a statistical mixture design. RSC Sustain. **2**(4), 975–987 (2024). https://doi.org/10.1039/D3SU00374D
59. King, O., Constant, E., Weems, A.C.: Shape memory poly(β-hydroxythioether) foams for oil remediation in aquatic environments. ACS Appl. Mater. Interfaces **13**(17), 20641–20652 (2021). https://doi.org/10.1021/acsami.1c02630
60. Wang, B., Karthikeyan, R., Lu, X.-Y., Xuan, J., Leung, M.K.H.: Hollow carbon fibers derived from natural cotton as effective sorbents for oil spill cleanup. Ind. Eng. Chem. Res. **52**(51), 18251–18261 (2013). https://doi.org/10.1021/ie402371n

61. Xu, M., et al.: Adsorption and foaming properties of edible egg yolk peptide nanoparticles: effect of particle aggregation. Curr. Res. Food Sci. **4**, 270–278 (2021). https://doi.org/10.1016/j.crfs.2021.04.002
62. Naser, D.M.I.: Expended polystyrene (EPS) waste as sorbent for crude oil spill cleanup: from laboratory experiments to field application. J. Pet. Res. Stud. **7**(1), 258–272 (2021). https://doi.org/10.52716/jprs.v7i1.180
63. Ali, S.: Synthesis and characteristics of oleophilic gel polymer for oil spills cleanup. Afr. J. Eng. Res. **7**(3), 85–93 (2019). https://doi.org/10.30918/AJER.73.19.032
64. Anwana, E.E., Ewemoje, O.E.: Oil sorption performance of sorbent materials examined under static and dynamic conditions. Eur. J. Eng. Technol. Res. **6**(5), 107–110 (2021). https://doi.org/10.24018/ejeng.2021.6.5.2490
65. Subramoniapillai, V., Thilagavathi, G.: Oil spill cleanup by natural fibers: a review. Res. J. Text. Appar. **26**(4), 390–404 (2022). https://doi.org/10.1108/RJTA-03-2021-0031
66. Ranganathan, P., Chen, C.-W., Tasi, M.-C., Rwei, S.-P., Lee, Y.-H.: Biomass thermoplastic (co)polyamide elastomers synthesized from a fatty dimer acid: a sustainable route toward a new era of uniform and bimodal foams. Ind. Eng. Chem. Res. **60**(33), 12139–12154 (2021). https://doi.org/10.1021/acs.iecr.1c01533
67. Huang, X.-Y., et al.: Thermal stability analysis of polyurethane foams made from microwave liquefaction bio-polyols with and without solid residue. BioResources **13**(2) (2018). https://doi.org/10.15376/biores.13.2.3346-3361
68. Valencia, L., Rosas, W., Aguilar-Sanchez, A., Mathew, A.P., Palmqvist, A.E.C.: Bio-based micro-/meso-/macroporous hybrid foams with ultrahigh zeolite loadings for selective capture of carbon dioxide. ACS Appl. Mater. Interfaces **11**(43), 40424–40431 (2019). https://doi.org/10.1021/acsami.9b11399
69. Członka, S., Kairytė, A., Miedzińska, K., Strąkowska, A.: Polyurethane composites reinforced with walnut shell filler treated with perlite, montmorillonite and halloysite. Int. J. Mol. Sci. **22**(14), 7304 (2021). https://doi.org/10.3390/ijms22147304
70. Kairytė, A., Członka, S., Boris, R., Vėjelis, S.: Vacuum-based impregnation of liquid glass into sunflower press cake particles and their use in bio-based rigid polyurethane foam. Materials (Basel) **14**(18), 5351 (2021). https://doi.org/10.3390/ma14185351
71. Abdullah, N.A., Hamdi, H.A., Dareb Shaban, A.: The effect of chicken Gallus gallus (Domestics) feathers on the sorption properties of polyurethane foam. Bionatura **7**(4), 1–5 (2022). https://doi.org/10.21931/RB/2022.07.04.63
72. García-Rengifo, A.R., Rojas-Bringas, P.M., De-la-Torre, G.E., Torres, F.G.: Environmental impact of peanut skin-reinforced native starch foams modified by acetylation. Environ. Qual. Manag. **31**(3), 89–99 (2022). https://doi.org/10.1002/tqem.21754
73. Członka, S., Strąkowska, A., Kairytė, A.: Application of walnut shells-derived biopolyol in the synthesis of rigid polyurethane foams. Materials (Basel) **13**(12), 2687 (2020). https://doi.org/10.3390/ma13122687
74. Carranza, I., et al.: Characterising and modelling the mechanical behaviour of polymeric foams under complex loading. J. Mater. Sci. **54**(16), 11328–11344 (2019). https://doi.org/10.1007/s10853-019-03673-8
75. Bolf, D., Zamarin, A., Krolo, P., Hadjina, M.: Experimental evaluation of shear properties of lightweight PVC core for marine application using digital image correlation system. J. Mar. Sci. Eng. **10**(2), 280 (2022). https://doi.org/10.3390/jmse10020280
76. Fathi, M., Sameezadeh, M., Vaseghi, M.: Compressive response and energy absorption of foam-filled aluminum honeycomb composite: experiments and simulation. J. Braz. Soc. Mech. Sci. Eng. **45**(11), 562 (2023). https://doi.org/10.1007/s40430-023-04489-z
77. Mazzuca, P., Firmo, J.P., Correia, J.R., Garrido, M.: Mechanical behaviour in shear and compression of polyurethane foam at elevated temperature. J. Sandw. Struct. Mater. **24**(2), 1429–1448 (2022). https://doi.org/10.1177/10996362211050919
78. Bailly, N., Petit, Y., Desrosier, J.-M., Laperriere, O., Langlois, S., Wagnac, E.: Strain rate dependent behavior of vinyl nitrile helmet foam in compression and combined compression and shear. Appl. Sci. **10**(22), 8286 (2020). https://doi.org/10.3390/app10228286

79. Ling, C., Ivens, J., Cardiff, P., Gilchrist, M.D.: Deformation response of EPS foam under combined compression-shear loading. Part I: experimental design and quasi-static tests. Int. J. Mech. Sci. **144**, 480–489 (2018). https://doi.org/10.1016/j.ijmecsci.2018.06.014
80. Viju, S., Thilagavathi, G., Keertana, N.: Oil sorption properties of cotton comber noil/recycled polyester blended needle punched nonwoven fabrics—an optimization study. J. Ind. Text. **52** (2022). https://doi.org/10.1177/15280837221113090
81. Sahu, B., Dolui, S., Maity, S., Gurajala, K.C., Banerjee, S.: Used cooking oil-derived fluorinated porous polysulfide materials for oil spill remediation: a waste-to-wealth by sustainable initiative. ACS Appl. Eng. Mater. **2**(7), 1866–1872 (2024). https://doi.org/10.1021/acsaenm.4c00289
82. Ghadei, S.K., Sankaran, K.J., Sakthivel, R.: Multifunctional potential of Dandelion-like structured foam for self-cleaning, oil/water separation, and wearable sensing. ACS Appl. Nano Mater. **7**(15), 17898–17912 (2024). https://doi.org/10.1021/acsanm.4c03084
83. Zimmermann, M.V.G., Zattera, A.J., Santana, R.M.C.: Sorbent system based on acetylated microfibrillated cellulose for remediation of oil aquatic environments. Matéria (Rio Janeiro) **24**(2) (2019). https://doi.org/10.1590/s1517-707620190002.0692
84. Mukherjee, M., Ghadei, S.K., Rakesh, B., Balaji, U., Sakthivel, R., Sankaran, K.J.: Fluorine-free sustainable three-dimensional superhydrophobic and superoleophilic robust foam for efficient oil/water separation. ACS Appl. Eng. Mater. **2**(2), 324–334 (2024). https://doi.org/10.1021/acsaenm.3c00626
85. Topuz, F., Abdulhamid, M.A., Nunes, S.P., Szekely, G.: Hierarchically porous electrospun nanofibrous mats produced from intrinsically microporous fluorinated polyimide for the removal of oils and non-polar solvents. Environ. Sci. Nano **7**(5), 1365–1372 (2020). https://doi.org/10.1039/D0EN00084A
86. Visco, A., Quattrocchi, A., Nocita, D., Montanini, R., Pistone, A.: Polyurethane foams loaded with carbon nanofibers for oil spill recovery: mechanical properties under fatigue conditions and selective absorption in oil/water mixtures. Nanomaterials **11**(3), 735 (2021). https://doi.org/10.3390/nano11030735
87. Piperopoulos, E., Calabrese, L., Mastronardo, E., Abdul Rahim, S.H., Proverbio, E., Milone, C.: Assessment of sorption kinetics of carbon nanotube-based composite foams for oil recovery application. J. Appl. Polym. Sci. **136**(14) (2019). https://doi.org/10.1002/app.47374
88. Calabrese, L., et al.: Sorption capacity of polydimethylsiloxane foams filled with thermal-treated bentonite—polydimethylsiloxane composite foams for oil spill remediation. Materials (Basel) **16**(13), 4818 (2023). https://doi.org/10.3390/ma16134818
89. Wang, X., Zhang, M., Schubert, D.W., Liu, X.: Oil–water separation polypropylene foam with advanced solvent-evaporation induced coexistence of microspheres and microporous structure. Macromol. Rapid Commun. **43**(17) (2022). https://doi.org/10.1002/marc.202200177
90. Cifarelli, A., Boggioni, L., Vignali, A., Tritto, I., Bertini, F., Losio, S.: Flexible polyurethane foams from epoxidized vegetable oils and a bio-based diisocyanate. Polymers (Basel) **13**(4), 612 (2021). https://doi.org/10.3390/polym13040612
91. Wang, J., Zhang, C., Deng, Y., Zhang, P.: A review of research on the effect of temperature on the properties of polyurethane foams. Polymers (Basel) **14**(21), 4586 (2022). https://doi.org/10.3390/polym14214586
92. D'Arienzo, L., Incarnato, L., Rinaldi, S., Scarfato, P.: Durability of a Soya-derived flexible PU foam for mattresses: effects of different artificial aging. Macromol. Symp. **405**(1) (2022). https://doi.org/10.1002/masy.202100247
93. Abd Rahim, N.S., Saaid, I.M., Umar, A.A.: Evaluation of foam performance at different temperature for enhanced oil recovery process. World J. Eng. **16**(3), 412–418 (2019). https://doi.org/10.1108/WJE-06-2018-0210
94. Obisesan, O., Ahmed, R., Amani, M.: The effect of salt on stability of aqueous foams. Energies **14**(2), 279 (2021). https://doi.org/10.3390/en14020279
95. Mi, H.-Y., et al.: Highly durable superhydrophobic polymer foams fabricated by extrusion and supercritical CO_2 foaming for selective oil absorption. ACS Appl. Mater. Interfaces **11**(7), 7479–7487 (2019). https://doi.org/10.1021/acsami.8b21858

96. Liang, M., Zhao, X., Wang, J., Feng, Y.: A comparative study on CO_2-switchable foams stabilized by C22- or C18-tailed tertiary amines. Molecules **28**(6), 2567 (2023). https://doi.org/10.3390/molecules28062567
97. Fang, Z., et al.: Development of high-performance biodegradable rigid polyurethane foams using full modified soy-based polyols. J. Agric. Food Chem. **67**(8), 2220–2226 (2019). https://doi.org/10.1021/acs.jafc.8b05342
98. Santos, O., Silva, M., Yoshida, M.: Synthesis and performance of different polyurethane foams as oil sorbents. J. Appl. Polym. Sci. **134**(41) (2017). https://doi.org/10.1002/app.45409
99. Muhammad, A., Neamh, H., Hameed, Z., Nasir, M.: Expanded polystyrene (EPS) waste as sorbent for crude oil spill cleanup: from laboratory experiments to field application. J. Pet. Res. Stud. **7**(1), 258–272 (2021). https://doi.org/10.52716/jprs.v7i1.180
100. Wang, S., et al.: Open-cell polypropylene/polyolefin elastomer blend foams fabricated for reusable oil-sorption materials. J. Appl. Polym. Sci. **133**(33) (2016). https://doi.org/10.1002/app.43812
101. Wu, M., et al.: Compressible carbon sponges from delignified wood for fast cleanup and enhanced recovery of crude oil spills by Joule heat and photothermal effect. Adv. Funct. Mater. **31**(3) (2021). https://doi.org/10.1002/adfm.202006806
102. Olszewski, A., et al.: Effect of bio-polyol molecular weight on the structure and properties of polyurethane–polyisocyanurate (PUR–PIR) foams. Sci. Rep. **14**(1) (2024). https://doi.org/10.1038/s41598-023-50764-3
103. Tomon, T.R.B., et al.: A novel naturally superoleophilic coconut oil-based foam with inherent hydrophobic properties for oil and grease sorption. Sci. Rep. **14**(1) (2024). https://doi.org/10.1038/s41598-024-64178-2
104. Piperopoulos, E., et al.: Bentonite–PDMS composite foams for oil spill recovery: sorption performance and kinetics. J. Appl. Polym. Sci. **139**(41) (2022). https://doi.org/10.1002/app.53003
105. Li, Z., et al.: Easily fabricated low-energy consumption Joule-heated superhydrophobic foam for fast cleanup of viscous crude oil spills. ACS Appl. Mater. Interfaces **13**(43), 51652–51660 (2021). https://doi.org/10.1021/acsami.1c13574
106. Zemła, M., Prociak, A., Michałowski, S.: Bio-based rigid polyurethane foams modified with phosphorus flame retardants. Polymers **14**(1), 102 (2021). https://doi.org/10.3390/polym14010102
107. Herren, B., Saha, M.C., Altan, M.C., Liu, Y.: Funnel-shaped floating vessel oil skimmer with Joule heating sorption functionality. Polymers **14**(11), 2269 (2022). https://doi.org/10.3390/polym14112269
108. Rizvi, A., Chu, R., Lee, J.H., Park, C.: Superhydrophobic and oleophilic open-cell foams from fibrillar blends of polypropylene and polytetrafluoroethylene. ACS Appl. Mater. Interfaces **6**(23), 21131–21140 (2014). https://doi.org/10.1021/am506006v
109. Al-Marri, M.J., Kuti, Y.O., Khraisheh, M., Kumar, A., Khader, M.M.: Kinetics of CO_2 adsorption/desorption of polyethyleneimine–mesoporous silica. Chem. Eng. Technol. **40**(10), 1802–1809 (2017). https://doi.org/10.1002/ceat.201600452
110. Singh, V., et al.: Novel natural sorbent for oil spill cleanup. Ind. Eng. Chem. Res. **53**(30), 11954–11961 (2014). https://doi.org/10.1021/ie5019436
111. Toteva, V., Staneva, D., Grabchev, I.: Pollutants sorbent made of cotton fabric modified with chitosan–glutaraldehyde and zinc oxide particles. Materials **14**(12), 3242 (2021). https://doi.org/10.3390/ma14123242
112. Omisol, C.J.M., et al.: Flexible polyurethane foams modified with novel coconut monoglycerides-based polyester polyols. ACS Omega **9**(4), 4497–4512 (2024). https://doi.org/10.1021/acsomega.3c07312
113. Oribayo, O., Pan, Q., Feng, X., Rempel, G.L.: Hydrophobic surface modification of FMSS and its application as effective sorbents for oil spill clean-ups and recovery. AIChE J. **63**(9), 4090–4102 (2017). https://doi.org/10.1002/aic.15767

Chapter 4
Properties of Bio-based Foam Sorbents

Abstract Bio-based foam sorbents require thorough characterization to understand their suitability for oil spill remediation. This chapter provides a detailed review of the various techniques used to evaluate the morphological, thermal, mechanical, and chemical properties of these materials. It covers spectroscopic techniques, such as Fourier Transform Infrared (FTIR) and Nuclear Magnetic Resonance (NMR) spectroscopy, as well as microscopy techniques like Scanning Electron Microscopy (SEM) and Field Emission Scanning Electron Microscopy (FESEM). The chapter also explores elemental composition analysis, thermal analysis, mechanical property testing, and surface and pore analysis. Techniques for measuring crystallinity and adsorption properties are also discussed. In addition to characterization methods, the chapter emphasizes key performance metrics, including oil sorption capacity, hydrophobicity, recyclability, and chemical stability. These metrics are crucial for assessing the effectiveness and durability of bio-based foam sorbents in practical applications. The chapter ultimately provides a comprehensive understanding of how these properties and performance indicators contribute to the successful use of bio-based foam sorbents in oil spill remediation.

Keywords Hydrophobicity · Foam morphology · Oleophilicity · Pore size · Fillers · Additives · Sorption capacity · Thermal stability · Flexibility · Rigid · Isotherms

Foam sorbents are versatile materials that can be applied for environmental remediation, mostly for oil spill recovery, wastewater treatment, or confinement of pollutants. These materials contain porous structures and surface properties that allow them to absorb and retain oils from aqueous environments. Foam sorbent development has become an interesting aspect globally because of the need to have environmental sustainability as well as effective means of industrial waste management and cleaning up of oil spills.

More importantly, the use of bio-based material in foam sorbent production makes it a more sustainable process. The renewable sources of foams from plant

A. A. Lubguban et al., *Bio-based Foam Sorbents*, Engineering Materials,
https://doi.org/10.1007/978-981-96-1460-8_4

oils, lignin, and other biomass sources provide an alternative that offers environmental friendliness, unlike traditional petrochemical-based foams. Apart from cutting down on dependency on non-renewable sources, the bio-based foams have specific characteristics that enhance their capacities for sorption, hydrophobicity, and oleophilicity.

Recent progress in the preparation of foam sorbents includes chemical modification and enhancement in structures. These modifications should increase both the efficiency of sorption and mechanical durability and reusability in foams. The pore structure, density, surface morphology, and functional additives seriously influence the performance of foam sorbents. This work has explored the properties and performance metrics of foam sorbents, particularly their wide range of applications together with bio-based materials offering a possibility of environmentally friendly clean-up solutions.

4.1 Morphology and Structure

4.1.1 Castor Oil-Based Foam Sorbents

Martins et al. [1] incorporated açaí seed biomass (5–20 wt%) into castor oil-based polyurethane, obtaining an eco-friendly sorbent for removal of waste cooking oil in food industry by-products. Based on the captured SEM images, the modified PU foam had a macroporous structure (pore size from 152 to 119 μm). The addition of açaí seed biomass reduced the pore size of the foams. Figure 4.1 demonstrates this reduction, leading to increased number of pores per area and apparent density. This behaviour could directly affect absorption capacity, increasing the amount of waste captured. Furthermore, açaí seed biomass during the polymerization process increased the nucleation sites for bubble formation but led to irregular cell structure. Larger fillers, such as açaí seed biomass, do not blend as homogenously as nanofillers. Hence, regions of increased nucleation sites are found within the matrix, leading to larger macropores than the rest of the material.

Another work by Martins et al. [2] modified castor oil-based PU foam using palm fiber residue. Much like adding açaí seed biomass, palm fiber residue increased the pore density of the PU sorbent. The fiber acted as a crosslinking agent and nucleation site for bubble (cell) formation. The increased fiber content led to a more homogenous porous structure, especially at smaller filler particle sizes. Open cells are also observed, which are essential for storing captured oil.

Maia et al. [3] incorporated coffee husks residue (CR) into PU foam for diesel S10 and S500 recovery. The neat PU foam had an open-cell structure with an average pore size of around 284.0 μm. Incorporating coffee husks residue resulted in partially closed-cell structure (as opposed to an open structure of the neat PU) and a reduction in average pore size (Fig. 4.3). The average pore size in PU composites for various coffee husks loading ranged from 192.3 to 226.2 μm, which indicates much

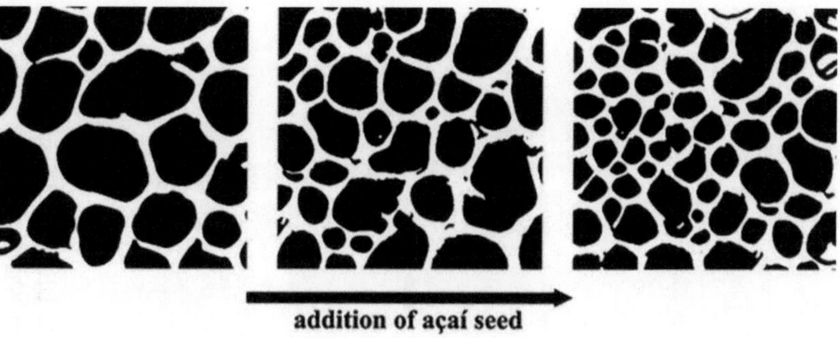

Fig. 4.1 Graphical representation of pore size reduction thru Açaí seed biomass addition

smaller cells and a denser foam structure. This shift toward a more closed-cell structure was commensurate with the increased density realized in the composite foams. Morphology was also less uniform and irregular as the CR content increased, with signs of increased surface roughness and cellular deformation.

The neat PU foam had a density of about 45.0 kg/m^3, close to the published values for these same types of materials. The addition of CR led to a gradual increase in the density of composite foams as the content of CR increased. For instance, while 5% CR added PU foams had a density of 47.0 kg/m^3, the density crept upward gradually to 83.5 kg/m^3 with PU + 40% CR. This can be attributed to the increase in viscosity effect during foam creation when higher contents of CR constrained the expansion of foam cells.

On the other hand, Perera et al. [4] functionalized castor oil-based PU foam with octadecyltrichlorosilane (C18)-modified diatomaceous earth particles. This modification showed remarkable changes in sorbent structure, improving oil sorption performance. Figure 4.2 shows the SEM images of the produced sorbents. The neat PU foam consisted of a smooth skeleton, having a very porous network. The addition of the C18-DE particles resulted in a rougher surface containing identifiable irregularities, increasing the overall surface roughness of the foam. The morphology became more heterogeneous on account of the particle distribution in the foam matrix. This surface modification enhanced the hydrophobicity and oleophilicity.

SEM images revealed the incorporation of C18-DE particles into the PU matrix and presented as embedded disk-like structures. The dispersion of the particles in the foam structure showed effective integration without significant aggregation, thus implying a homogenous modification at the surface level along the entire foam. The uniformity of dispersion contributed to an improved hydrophobic character with an enhanced hierarchical surface structure.

The modification also impacted the pore structure of the foam. The unmodified PU showed a relatively open-cell structure, while the pore structure for the modified PU showed partially closed-cell characteristics because of the presence of the C18-DE particles. The addition of C18-DE not only proved to be a cause of the rise of the surface roughness but also led to the development of smaller

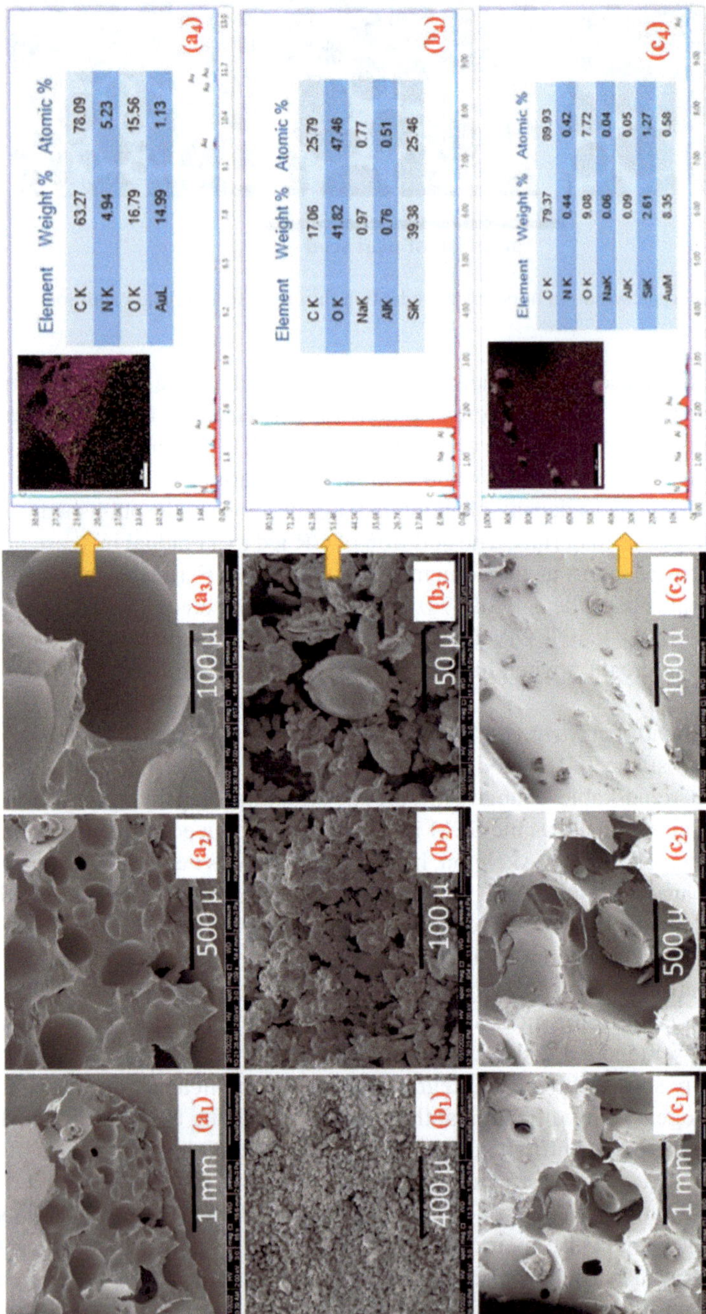

Fig. 4.2 SEM images on different scales: (a1–a3) pristine PU synthesized foam, (b1–b3) C18-DE, (c1–c3) modified PU-C18-DE foam, EDX graph mapping and elemental analysis for all three samples, respectively, (a4), (b4), (c4); By Perera et al. [4], licensed under CC BY 4.0. *No changes were made*

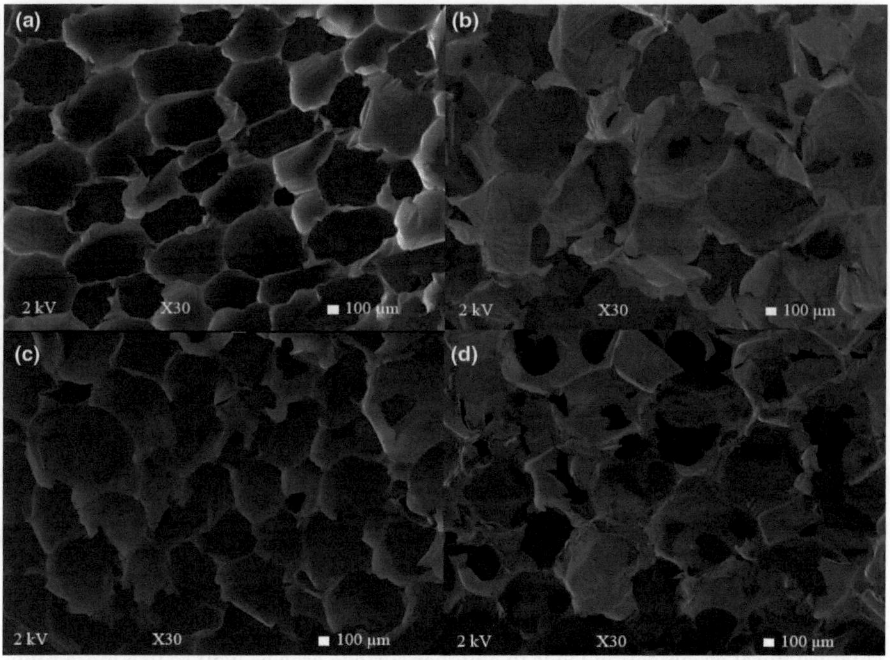

Fig. 4.3 Morphological Properties of POBP PU foam; **a** commercial PU, **b** POBP PU at 25 wt%, **c** POBP PU at 50 wt%, **d** POBP PU at 75 wt%. Reprinted with permission from Sittinun et al. [5]. Copyright ©Springer Science + Business Media, LLC, part of Springer Nature 2020. Licensed through the Copyright Clearance Center Marketplace

pores. Such changes are attributed to the nucleation effect of the particles themselves during the foaming process, causing a more closed-cell, denser architecture, which is particularly favorable for oil sorption.

4.1.2 Other Plant-Oil Seed Based Foam Sorbents

Sittinun et al. [5] successfully synthesized palm-oil-based polyol (POBP) for the production of polyurethane foam sorbents. Scanning electron microscopy revealed that a higher percentage of POBP causes the foam to shift from mostly closed-cell to mostly open, interconnected morphologies, indicating pore dynamics and possibly their capacities for sorption due to increased surface areas.

Polyols typically derived from petrochemicals lead to the formation of fairly uniform closed cells. POBP introduces a more heterogeneous morphology. An increasing content of POBP leads to a very drastic increase in pore size and openness of the cell configuration. This is mainly due to the much lower reactivity of POBP compared to the traditional petrochemical polyols. This decreases the reactivity of

POBP and slows down the foaming reaction, thereby allowing for the formation of larger, more interconnected pores.

The transition from a closed to an open-cell morphology of POBP-based polyurethane foams can be further explained by the effects that POBP exert on polymer network formation. POBP incorporates unsaturated fatty acids into the polyol blend that are structurally less rigid than petrochemical polyols merely due to the presence of double bonds in their fatty acid chains. This gives them elasticity that prohibits the uniform cell structure typically seen with conventional foams. The mechanism is likely to be explained by a synergy between reduced hydrogen bonding, and cross-linking density are critical factors for cellular rigidity in polyurethane foams. Besides, the incorporation of hydroxyls in POBP increases intermolecular interaction and hence has a higher pore connectivity by allowing a cell opening during foam expansion (Fig. 4.3).

Meanwhile, Tomon et al. [6] utilized a bio-based polyol from coconut oil for a superoleophilic and inherently hydrophobic sorbent. The authors found that the naturally superoleophilic foam (NSF) has a cell structure that is more porous and homogenously distributed compared to the control petroleum-based foam having larger irregular cells.

Figure 4.4 shows the SEM images of the sorbent. The defined cell arrangement in NSF enhances its oil sorption ability because oil can be retained more effectively. This structural organization leads to a higher surface area as well as to a more robust three-dimensional network suitable for oil retention.

The improved NSF morphology is attributed to the presence of polyol-based coconut oil. Due to this, there is more cross-linking density in the foam matrix as it has a higher hydroxyl (OH) number. Excess hydroxyl groups are favorable for the formation of urethane linkages, thus stiffening the foams' structure. The polyester-based polyol of coconut oil stabilizes cell walls of the foam, thus increasing their mechanical properties. This subsequently leads to uniform cell size distribution and

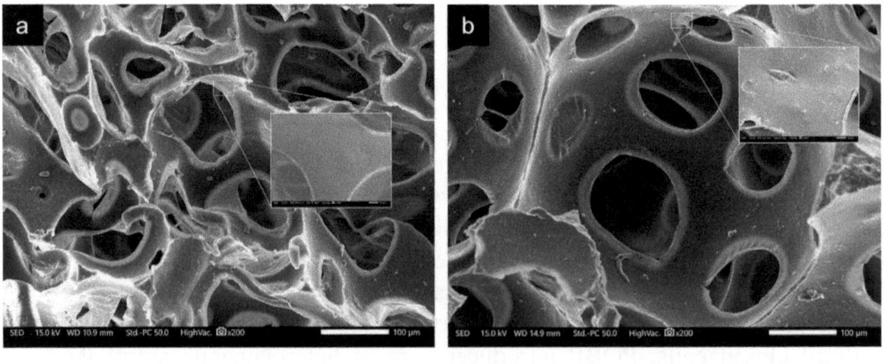

Fig. 4.4 SEM images of the **a** control polyurethane foam and **b** naturally superoleophilic foam (NSF) at × 200 with × 1000 magnification. By Tomon et al. [6], licensed under CC BY 4.0. *No changes were made*

enhances its macro-porous network, leading to exceptional oil sorption capacity and hydrophobic properties.

Among the morphological features characterizing NSF, a rough surface is essential for the hydrophobic nature. The surface irregularity from the SEM images could be associated with the fatty acid chain from the coconut oil-derived polyol, thus creating micro-scale surface irregularities. These micro-scale surface irregularities trap air pockets inside the foam structure hence aid in improving hydrophobicity. The rough surface serves as a barrier against penetrating water; this further combines with the hydrophobic nature of the foam in making it highly suitable for applications that separate oil from water.

While there are morphological differences, the average open-cell content of NSF is still similar to that of the control foam, only at 97.21 versus 97.16%, but the inclusion of coconut oil-based polyol transformed the internal cell architecture, resulting in increased porosity and connectivity of cells for higher fluid flow and absorption of oil-based contaminant. This thus indicates that while open cell content is an important parameter, uniformity and connectivity of the internal structure is of the same importance in determining foam sorbent performance.

Another work by Cuong et al. [7] utilized a sunflower-oil based polyol for petro-polyol replacement and synthesized PU foam sorbents (SC-PU) with different loadings of sugarcane bagasse as fillers. A half-open cell structure is formed in SC-PU composites, while the neat PU assumes a closed-cell nature. Open-cell morphology ensures greater surface area and more extensive capillary networks for effective oil absorption.

The incorporation of sugarcane bagasse into a polyurethane matrix introduces a hierarchical porous structure. During polymerization, the fibrous structure of the bagasse filler interferes with the homogeneity of the polyurethane matrix and therefore enhances the creation of the interconnected network of pores. The cellulose content of sugarcane bagasse is very high, about 43.2%, which also enhances the porosity of the composite. This increased area serves not only for interaction with oil but also for enhanced mechanical integrity of the foam, thanks to fiber reinforcement; hence, it retains more oil.

The sunflower oil-based polyol, with a hydroxyl number of approximately 182 mg KOH/g, introduces bio-functional groups interacting with the lignocellulosic filler. The flexibility of the bio-polyol used here, compared to the more rigid petrochemical polyols, enables better compatibility with the sugarcane fibers, therefore favoring a well-distributed porosity in the foam matrix. On its turn, the hydroxyl functionality, being larger (3.5 OH/molecular groups), increases the cross-linking density, hence stabilizing the open-cell structure.

4.1.3 Lignin-Based Foam Sorbents

In the studies found in the literature, the porous morphology of the lignin-based polyurethane foams (LPUFs) was a crucial factor for oil adsorption. Indeed, the more interconnected and homogeneous structure of pores allowed higher oil absorption and retention while maintaining integrity for structural purposes. Surface modification of the foams, either with nanoparticles or chemical treatments, substantially enhanced the functional performance of the foams, depending on the application. The increased surface roughness enhanced hydrophobicity or hydrophilicity and improved the efficiency of oil–water separation.

Mohammadpour and Sadeghi [8] studied the effect of liquefied lignin content on the synthesis of bio-based PU foam for oil adsorption application. The morphological analysis performed by the authors revealed changes as a function of increasing lignin content, from 25 to 100%, and significant modifications in the microstructure, porosity, and their resulting oil absorption capacity (Fig. 4.5).

With a 25% lignin content, the SEM images (4.5) showed a well-ordered open-cell structure with small and evenly distributed pores. Thin cell walls and uniform

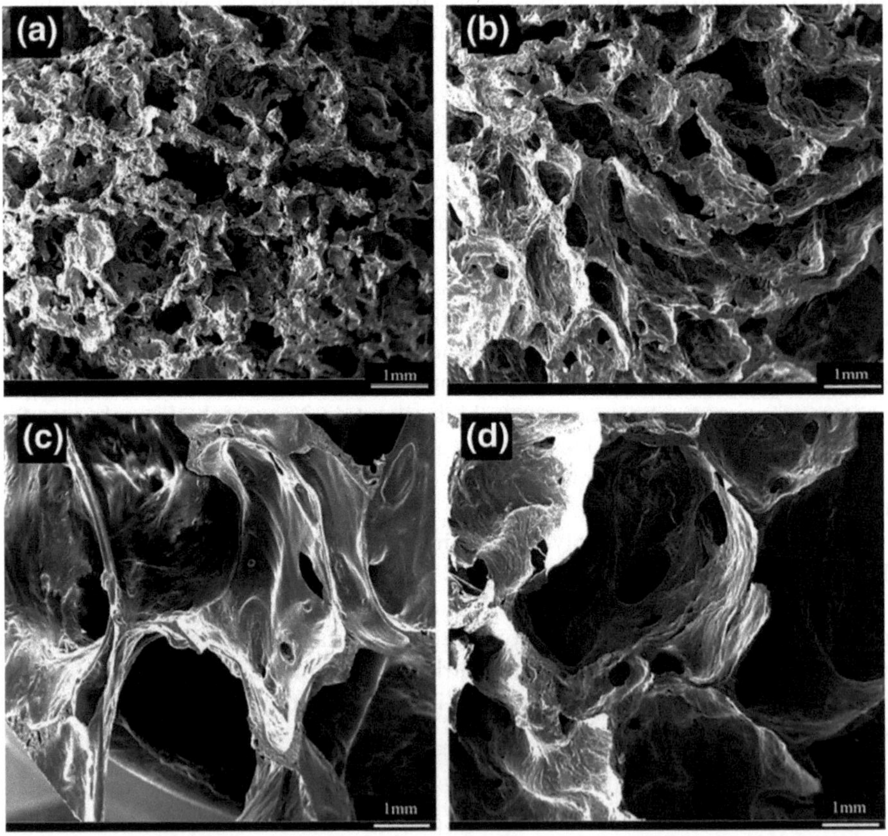

Fig. 4.5 Morphological Analysis of LPUFs at increasing lignin content; **a** 25%, **b** 50%, **c** 75%, **d** 100%. Reprinted with permission from Mohammadpour and Sadeghi [8]. Copyright ©Springer Science + Business Media, LLC, part of Springer Nature 2020. Licensed through the Copyright Clearance Center Marketplace

pores contributed to a highly porous and flexible character of the foam matrix. This morphology is characteristic of open-cell flexible polyurethane foams, of which the prevalence of PPG guarantees that the foam is elastic and thus conducive to oil absorption. The high surface area resulting from this regular distribution of pores gave it especially high performance in the uptake of oil, as the oil permeated and filled up the porous structure readily.

As the lignin content increased (50–75%), the morphology of the foam became irregular and disordered. The pore sizes increased and were more strongly varied; open and closed cells can be seen. Thicker cell walls are a result of the enhanced rigidity due to the relevant molecular architecture of lignin, which adds to the stiffness of the matrix of the foam. Although the pores were still interconnected, this variability in size and shape caused a slight collapse in certain areas, reducing the elasticity in the foam. Increased rigidity and a less homogeneous structure started to negatively impact oil absorption efficiency due to the incomplete diffusive path of oil through the irregular pore network within the foam.

With high lignin content (100%), the microstructure of the foam became highly irregular, with large and irregularly shaped pores and a very significant number of closed cells. The pore walls significantly thickened, and the structure looked much denser and more prone to collapse. This structural collapse, along with thick, closed-cell walls, significantly reduced the foam porosity and surface area, hence hampering oil absorption. The restricted flexibility and reactivity of lignin at high concentration depressed the foam matrix to a much less effective means for oil diffusion due to the collapsed and closed pores that reduced oil from penetrating and being absorbed efficiently.

Ma et al. [9] modified lignin-based PU foam with carbon nanotubes (CNT) to synthesize a photothermal sorbent for heavy oil adsorption. Scanning electron microscopy showed a very porous and, more importantly, an interconnected cellular structure. The LPUF samples exhibited high porosity with the form of three-dimensional interconnected structures. The foam exhibited open-cell morphology with pores 200–1000 μm in pore diameter and varied as a function of CNTs concentration within the foam matrix. This distinctive open-cell structure is very important for oil absorption; namely, it can allow easy oil infiltration and surface contact maximization.

Pores sizes were increased with CNTs concentration, as illustrated samples of 3% CNT show higher cell sizes due to increased viscosity and lesser reaction rate of foaming. However, the addition of CNT greatly changed pore size and distribution for the LPUF. For samples containing 0.5% and 1%, the pore size was uniform at about 300 μm; however, for all the samples containing CNT levels $\geq$ 1.5%, the cell size became larger with an reaching the size of 1000 μm (at 3%). Moreover, the embedding of CNTs into foam walls enhances the surface roughness of LPUF and ensures hydrophobic properties.

Chen et al. [10] modified the surface of LPUF using polydopamine (PDA) and phytic acid (PA) to produce a multifunctional superhydrophilic/underwater superoleophobic sorbent for water/oil separation and water purification. PDA was deposited on the foam's surface, followed by modification with PA. These steps were very important for creating the required surface roughness and enhancing the foam's wettability properties. In the present study, SEM provided detailed pictures of the surface of the foam, showing that in-situ polymerization resulted in a highly roughened structure of the PDA particles. The nano- to micro-scale roughening increased the effective surface area of the foam, enhancing its wettability properties.

The deposition of polydopamine particles resulted in a finely textured surface, which was responsible for the increase in surface roughness. Such roughness contributes to superhydrophilicity as it allows water to spread fast across the surface of the foam and be absorbed. Such a texture, comprising PDA nanoclusters, increased nucleation points within the substrate for its water adhesion, increasing the permeation of water through the foam.

The following coating of phytic acid further modified the surface morphology, making it more hydrophilic. PA introduced phosphate groups, and this chemically improved the interaction of water with the surface of the foam. From SEM images, the PA coating was found to be well distributed on the surface and added a second layer of roughness contributing to the superoleophobic behavior of the foam underwater.

The same was observed by Wu et al. [11]. Authors added 1H, 1H, 2H, 2H-perfluorododecyl trichlorosilane (F-SiO_2) to LPUF to produce a superhydrophobic PU foam for oil–water separation. SEM analysis was highly important for determining the microstructure of the F-SiO_2 nanoparticles-based lignin polyurethane foam. The characteristics of the foam produced are highly porous, with an open-cell structure and pore sizes of about 200–500 μm, which significantly improve both oil absorption and mechanical flexibility. The open and interconnected pores allow for easy penetration and retention of oil, optimizing the foam's adsorption capacity while maintaining its lightweight characteristics, ideal for large-scale oil remediation.

The distribution of SiO_2 nanoparticles has been found to be quite uniform over the foam matrix, where the particles embed on the foam surface and inside the pore walls. Spherical nanoparticles measure approximately 480 nm, contributing to high surface roughness. This hierarchical roughness at the nano- to micro-scale significantly enhanced the foam's superhydrophobicity. Nanoscale roughness provided by SiO_2 nanoparticles increased the water contact angle to 151.3°, reducing the effective contact area of water with the foam surface, which is crucial for oil–water separation. The foam selectively absorbed oil while repelling water.

The microscale roughness of the foam, driven by its cellular architecture, increased the likelihood of trapping air pockets when submerged in water. The undulating structure of the foam walls, coated with F-SiO_2 nanoparticles, reduced water wetting and reinforced superhydrophobic properties. This combination of nano- and micro-scale roughness provided the necessary foundation for the high-performance separation of oil from water and demonstrated excellent performance in oil spill cleanups and water purification.

4.1.4 PLA-Based Foam Sorbents

Li et al. [12] explored the use of polylactic acid for foam sorbents using the supercritical CO2 foaming process. However, the synthesis of high-performing sorbents through this route faces challenges such as producing high-porosity and open-cell PLA foams. The formation of an open-cell structure puts very tight demands on the melt strength: high melt strength is needed to prevent excessive gas release, cell collapse, and low porosity; however, a lower melt strength is required for an open-cell structure. Thus, it is rather tricky to balance the melt strength since poor cell formation results from inadequate strength, while too high a strength counteracts the desired cell opening.

Authors added poly(butylenes succinate) (PBS) to the polymer matrix to address the issues. Based on the SEM micrographs, distinct differences are found in the PLA/PBS foams depending on composition. PLA/PBS foams showed a highly interconnected lace-like open-cell structure. The amount of open-cell content was as high as 98.2%, which accounted for the efficient oil adsorption. PLA/PBS blends displayed a dual-phase structure; in this case, the PBS phase played the role of a cell opener to decrease the crystallization of the PLA phase effectively. While at high foaming temperatures (115 °C), cell sizes were uniformly distributed between 100–300 μm, lower temperatures of 105 °C resulted in the formation of spherulites in the PLA/PBS0 foam, interfering with the growth of cells.

The addition of PBS enhanced the open-cell morphology. As the PBS content increased, a heterogeneous melt structure favored cell wall rupture, resulting in a higher degree of cell opening. For example, PLA/PBS2 foam demonstrated a very high expansion ratio (43.6) and porosity (97.7%), outperforming other reported PLA-based foams. However, an excessive PBS content (40–50 wt%) led to coarsening of the phase morphology, causing irregular cell shapes and reduced expansion ratios.

Temperature has been an important factor in defining cell structure. Most of the foams at 115 °C showed an amorphous character, which facilitates uniform cells. On the contrary, partial crystallization at lower foaming temperatures made cell nucleation difficult and reduced foam expansion.

Another work on supercritical CO_2 foaming was done by Yang et al. [13]. Authors utilized polylactic acid (PLA) and poly (3-hydroxybutyrate-co-3-hydroxyvalerate) (PHBV) for bio-based biodegradable polymer foam sorbents for oil–water separation. Compared with the pure PLA foams, the morphology of PLA/PHBV composite foams is significantly enhanced. Pure PLA foams fabricated through supercritical CO_2 foaming have large, closed-cell structures with a volume expansion ratio (VER) of 80.08. On the other hand, the addition of PHBV leads to a more open-cell structure with better cell connectivity, enhancing porosity. Open cell content (OCC) increases from 16.56 to 62.37% at 1 wt% PHBV, and 95.28% at 10 wt% PHBV. However, the VER decreases with PHBV addition, although it is still better than those found in the literature. The average cell size also decreases, thus increasing the cell density. Certainly, PHBV enhances the foaming of PLA. The entanglement of PLA chains through PHBV is crucial in bubble nucleation and growth of cell. Without PHBV, pure PLA has poor crystallization under 180 °C, resulting in closed-cell structures due to the lack of hard regions for a heterogeneous melt.

4.1.5 Other Biomass-Based Foam Sorbents

Tan et al. [14] synthesized a honeycomb-like foams from larch sawdust for sorbents in oil spills cleanup and recovery. The authors produced two types of these: liquefied-larch-based polymer foam (LLB-PF) and its carbonized product, liquefied-larch-based carbon foam (LLB-CF). SEM micrographs showed that LLB-PF and LLB-CF have three-dimensional, open-cell structures with interconnected honeycomb-like geometries. These features are highly desirable for oil and organic solvent absorption due to the large surface areas and numerous interconnected pores that allow for the diffusion and storage of absorbed liquids. The cell sizes for the polymer foam, LLB-PF, ranged from 300 to 450 μm, with pore sizes between 1.0 and 3.5 μm. After carbonization, the cell sizes in LLB-CF were reduced to 225–375 μm, while the honeycomb structure remained intact.

At higher magnifications, SEM and TEM images showed that the cell walls of both LLB-PF and LLB-CF were very smooth, which enhanced the foam's hydrophobicity. The foams also contained numerous circular pores distributed on the cell walls, connecting adjacent cells. This connectivity facilitated capillary action, allowing rapid oil absorption into the foam structure.

Carbonization had a great effect on the microstructure of the foam. While the total morphology of the foams remained after pyrolysis, there was volume shrinkage for the LLB-CF foam due to the volatilization of unstable polymer components within the foam. However, the structure composed of interconnected pores remained intact, and the density of LLB-CF (~ 9.7 mg/cm^3) was relatively low compared to that of LLB-PF (~ 13.5 mg/cm^3). This facilitated an improvement in the absorption capacity for oils and organic solvents.

Songsaeng et al. [15] fabricated composite foam sorbents using natural rubber latex (NR) and reduced graphene oxide (rGO) as an additive. The SEM images (Fig. 4.6) revealed that both pure natural rubber (NR) foam and NR/rGO composite foams exhibited open-cell structures with spherical-shaped pores. However, the cell size and morphology were notably influenced by the rGO content. As the rGO content increased, the foam's cellular structure became more interconnected, with larger pore sizes. This larger pore size enhanced the foam's oil adsorption capacity by increasing the surface area available for oil retention.

Increasing the rGO content from 0.25 to 1.5 phr led to a gradual increase in cell size and surface roughness. The NRG-0.5 sample demonstrated the most efficient oil adsorption behavior, featuring well-formed open pores and minimal rGO aggregation. However, higher rGO content (1.0 and 1.5 phr) resulted in larger, less uniform pores and a reduction in adsorption capacity, as the larger pores were less effective at retaining oil.

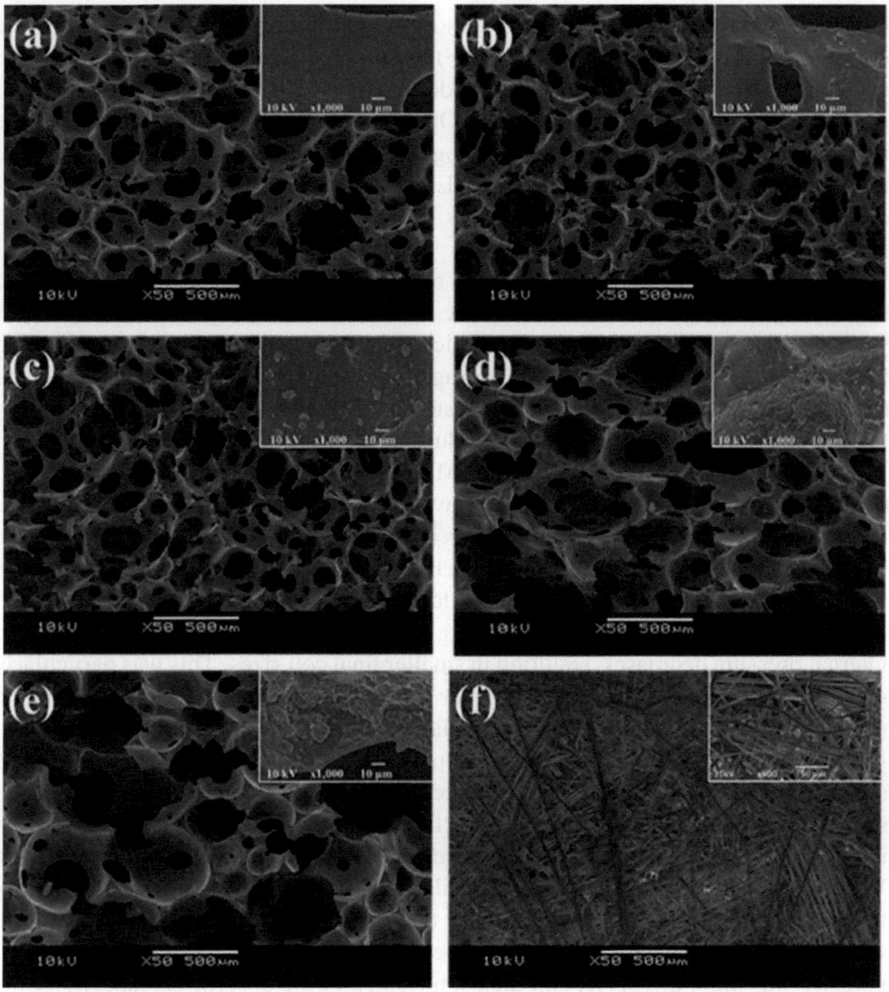

Fig. 4.6 SEM images (50 magnification) of **a** NR, **b** NRG-0.25, **c** NRG-0.5, **d** NRG-1.0, **e** NRG-1.5, and **f** CM sorbents. By Songsaeng et al. [15], licensed under CC BY 4.0. *No changes were made*

Some degree of aggregation or agglomeration of rGO within the NR matrix was observed in the samples. This aggregation became more pronounced at higher rGO loadings, particularly in the NR/rGO-1.0 and NR/rGO-1.5 samples. The rGO particles served as nucleation sites during the foaming process, resulting in irregular pore sizes and structures. Despite this, the addition of 0.5 phr rGO (NRG-0.5) demonstrated an optimal balance between pore structure uniformity and enhanced oil adsorption performance.

Zhang et al. [16] synthesized tannin-based polyether polyol (MTEBP) for the production of tannin extract-based foams (MTPFs). The morphological analysis indicated a noticeably different cellular morphology among the MTPF samples. The reference rigid polyurethane foam (RPUF) had an average cell size of 0.61 mm, with cells ranging from 0.31 to 1.30 mm. The addition of MTEBP to the foam matrix led to observable changes in cell morphology. On average, the cell size increased with the rise in MTEBP content, and a more homogeneous distribution of cells was observed in samples from MTPF10 to MTPF50. SEM images revealed well-developed open-cell structures with interconnected pores, which are crucial for enhancing oil and solvent absorption capacity.

With the addition of MTEBP in the range of 10–50 wt%, cell size increased slightly compared to RPUF. For instance, cells within the range of 0.60–0.79 mm were obtained for MTPF10-MTPF50, which was larger than those in RPUF, allowing for better pore connectivity and oil storage. This increase in cell size is attributed to the lower viscosity of the reactant system as MTEBP content increased, which favored the formation of larger bubbles during the foaming process.

The MTPF100 foam, in which 100% MTEBP was used as the polyol component, exhibited a reduction in cell size to an average of 0.60 mm, similar to that of the reference foam. The reduced cell size was linked to the faster gelling process caused by the increased water content in the MTPF100 system during foaming. Water reacts with isocyanate to release CO_2, which leads to rapid foam expansion. However, if the CO_2 diffusion rate surpasses the air diffusion rate into the foam cells, the internal pressure decreases, resulting in smaller final cell sizes. This was particularly noticeable in MTPF100, where the higher water content and heat release during the reaction accelerated cross-linking and gelation [16].

Wang et al. [17] worked on a novel composite foam sorbent using a combination of alginate and a hydroxyl aluminum stearate oil gelator. The authors aimed to synthesize a materials that could absorb oil and effectively retaining it, addressing common issues such as re-bleeding which has been observed in traditional oil absorbents. When probed with SEM, the alginate-oil gelator composite foam showed highly interconnected porosity characteristic of the alginate matrix. The pores showed a wide-range size distribution from nanometers to hundreds of micrometers. Overall porosity was high, at 98.4%, and thus contributed to the absorption capacity of the composite foam. The bulk density of the foam measured at 21.1 mg/cm^3 meant that handling it was relatively lightweight and ultimately improved oil absorption performance.

Micrographs confirmed the surface bonding of hydroxyl aluminum stearate oil gelator particles on the composite foam. Gelator particles were uniformly dispersed throughout the foam porous structure, sticking to the skeleton of the foam. Such gelator particles enabled the foam to have significantly increased hydrophobicity and oleophilicity, thus making it optimally wettable by, as well as able to absorb and retain oil. Elemental analysis validated the presence of important elements such as carbon (C), oxygen (O), sodium (Na), and aluminum (Al) on the surface of the foam, thereby showing the good incorporation of the oil gelator.

It has been found that through the introduction of t-butanol in the preparation process, the pore structure was further optimized into a multiscale hierarchical network. At a 37.5% t-butanol content, the foam had a well-developed fine pore network, where small pores form on the pore walls of larger ones. This hierarchical structure further increased the surface area up to 33.65 m(2)/g and also enhanced oil absorption efficiency through more active sites for capturing oil. The particular surface area along with the particle of gelators allowed the foam to selectively absorb oils without water ingress.

The porosimetry curves showed very wide pore size ranges from nanometers to micrometers. Average pore diameter was approximately 2.75 μm. The SEM images highlighted that macropores and mesopores were present in the foam matrix. Due to this wide range in pore sizes, the pores would accept both low and high viscosity oils. The high porosity, coupled with a hierarchical pore structure, allowed the structure to absorb oil quickly, but it also stored large amounts of absorbed oils.

Ma et al. [18] worked on a polydimethylsiloxane (PDMS) modified biomass carbon fibers foam immobilized with Ag_2O@Ag nanoparticles via the self-assembly, freeze-drying, carbonization, and hydrophobic modification processes (termed as PAAC). Authors utilized sisal leaves as the cellulose source for a photocatalytically active foam sorbent with application for cleanup of high-viscosity crude oil spills, oil–water separation and aqueous organic pollutant elimination. Based on the field emission scanning electron microscopy, The foam (PAAC) had a highly porous, three-dimensional interconnected structure, which is crucial for oil absorption. This structure provides a large surface area and enhances capillary action, enabling rapid oil uptake. The original cellulose fibers were pyrolyzed into carbon fibers (CFs), preserving their cellulose fiber structure but diameter reduction was observed. After functionalization with Ag_2O@Ag nanoparticles, the carbon fibers developed a rougher texture, as the nanoparticles (50–300 nm) coated the surface. This increased surface roughness further enhanced the foam's hydrophobicity and oil absorption capacity.

The FESEM micrographs further confirmed the uniform distribution of Ag_2O@Ag nanoparticles across the carbon fibers, enhancing the foam's photothermal properties. These nanoparticles enabled the foam to absorb sunlight efficiently, converting solar energy into heat. The resulting increase in surface temperature allowed the foam to quickly reduce the viscosity of highly viscous crude oil, thereby improving its absorption kinetics and facilitating faster oil uptake.

4.2 Sorption Capacity

Sorption capacity is a crucial measure of how effective foams are at cleaning up oil spills and helping with environmental remediation. Expressed in grams of oil absorbed per gram of foam (g/g) indicates how well a foam can take in and hold oil contaminants from water or soil. Factors such as the type of bio-based material used, the foam's porosity, its surface chemistry, and any chemical modifications that enhance oil attraction all influence sorption capacity. Understanding these elements is essential for designing foams with superior oil-absorbing performance tailored to specific environmental challenges.

Incorporating bio-based materials into foams makes them more environmentally friendly and improves their ability to absorb oil. Natural polymers like lignin, polylactic acid (PLA), plant seed oils, and biomass oils add functional groups and structural features that increase the foam's affinity for oil molecules. Enhancements like adding nanoparticles or natural fibers can further boost porosity and surface characteristics, leading to higher sorption capacities. The following sections will explore how different bio-based foams achieve enhanced sorption capacities through their unique compositions and structures.

4.2.1 Castor Oil-Based Foam Sorbent

Castor oil-based sorbent foams are exceptionally good at absorbing oils, thanks to the unique chemical makeup of castor oil and how it interacts during the foam's creation. Castor oil is rich in ricinoleic acid, which contains hydroxyl groups. These groups form strong urethane bonds with isocyanates, creating a robust yet flexible matrix that maintains an open-cell structure. This open structure is crucial because it allows oil to seep into the foam's pores easily. The hydroxyl groups make the foam hydrophobic, meaning it repels water but attracts oil—an essential feature for selectively soaking up oils while leaving water behind [4].

The foam's open-cell design offers a vast internal surface area for oil absorption. For example, when foams are reinforced with palm fiber residues, they can absorb up to 24 g of crude oil per gram of foam. This high capacity is due to the well-defined pores that let oil diffuse deep into the foam [2]. We can boost this sorption capacity even further with surface modifications. Adding hydrophobic fillers like octadecylsilane-modified diatomite or silicon carbide (SiC) nanoparticles increases both the surface area and hydrophobicity. Castor oil-based foams enhanced with modified diatomite showed a nearly threefold increase in oil absorption compared to unmodified foams [4].

Castor oil's flexibility as a polyol also enhances the foam's performance. Its fatty acid chains allow the foam to bend and stretch, accommodating large amounts of oil without breaking apart. This flexibility helps the foam quickly bounce back after being squeezed, making it ideal for repeated cycles of absorbing and releasing oil.

Incorporating agricultural residues like coffee husks can further improve sorption capacity by adding more tiny and large pores, which increase the available surface area. Foams containing 40% coffee husk residues showed over 80% efficiency in absorbing diesel oils [3]. By combining the natural properties of castor oil with added porosity, these foams can soak up large quantities of oil while staying stable.

Moreover, castor oil-based foams are highly oleophilic—they have a strong attraction to oil. They actively draw in and bind oil molecules even when water is present. For instance, foams used to absorb waste cooking oil reached capacities of 10.42 g of oil per gram of foam under stable conditions [1]. This oil-loving nature and their water-repellent surfaces make them highly effective for cleaning up oil spills.

The impressive sorption capacity of castor oil-based foams comes from their chemical structure, open-cell design, surface enhancements, and the addition of natural fillers. These foams provide sustainable and highly effective alternatives to traditional petroleum-based sorbents in efforts to remediate environmental pollution.

4.2.2 Other Plant Seed Oil-Based Foam Sorbents

Plant seed oil-based foam sorbents—derived from oils like coconut, palm olein, and sunflower—are exceptionally effective at absorbing oils. This efficiency is due to their hydrophobic (water-repelling) and oleophilic (oil-attracting) surfaces, porous structures, and unique chemical compositions. The fatty acids in these plant seed oils naturally create hydrophobic surfaces when used as polyols in foam production.

For example, coconut oil-based foam sorbents have demonstrated oil sorption capacities ranging from 14.89 to 24.65 g per gram of foam, depending on the type of oil involved [6]. This high capacity is attributed to the foam's superoleophobic surface, which efficiently absorbs oil while repelling water. The saturated fatty acids in coconut oil enhance the foam's ability to selectively soak up oil, making it ideal for separating oil from water.

The porous structure of these foams significantly boosts their sorption capacity. Open-cell designs provide a large surface area for oil interaction and absorption. Palm olein-based foam sorbents, for instance, feature an interconnected pore network that facilitates oil penetration deep into the foam matrix [5]. This increased contact area between the oil and the sorbent material leads to higher oil uptake.

Chemical composition also plays a crucial role, particularly the presence of hydroxyl groups in bio-based polyols. Sunflower oil-based polyols have a hydroxyl number that allows effective cross-linking during foam synthesis, resulting in a stable foam structure that enhances oil absorption [7]. The formation of urethane linkages further stabilizes the foam, contributing to its high sorption capacity.

Surface morphology—specifically rough textures—increases the available surface area for oil adsorption. Coconut oil-based foams exhibit a macro-porous structure with rough surfaces, which improves oil retention by trapping oil molecules within the foam matrix [6]. This combination of rough morphology and hydrophobicity ensures quick and efficient oil absorption.

Natural reinforcements like sugarcane bagasse can enhance sorption capacity and mechanical strength. Sunflower oil-based foam sorbents infused with sugarcane bagasse have absorbed up to 15.2 g of oil per gram of foam, nearly four times higher than foams without this filler [7]. The synergistic effect of bio-polyols and natural fillers creates a highly absorbent and environmentally friendly material.

The high sorption capacity of plant seed oil-based foam sorbents is driven by their hydrophobic and oleophilic surface chemistry, porous structures, and beneficial chemical compositions. Their recyclability and sustainability make them superior alternatives to conventional petroleum-based sorbents for oil spill cleanup and various industrial applications.

4.2.3 Lignin-Based Foam Sorbents

Lignin-based foams exhibit remarkable sorption capacities, primarily due to the unique chemical structure of lignin and its compatibility with the matrix. Lignin, a naturally occurring aromatic biopolymer, is rich in functional groups such as hydroxyl (–OH), methoxy ($–OCH_3$), and carboxyl (–COOH) groups. These functional groups enhance the foam's interaction with hydrophobic substances like oil, increasing the overall sorption efficiency. The presence of phenolic hydroxyl groups allows for hydrogen bonding and other interactions that strengthen the adsorption of oil molecules onto the foam surface [8].

The complex aromatic structure of lignin contributes to an increased surface area and porosity within the foam. This porous architecture provides more active sites for adsorption, allowing oil molecules to diffuse into the foam's interior and maximize the available adsorption surface. Wu et al. [11] reported that lignin-based foams exhibit high porosity, significantly enhancing their capacity to adsorb oils and organic contaminants.

Moreover, lignin's integration into the matrix improves the foam's structural properties. Incorporating lignin increases cross-linking density, leading to a more robust and thermally stable material. This structural enhancement ensures the foam maintains its integrity during sorption, preventing collapse and allowing for higher oil uptake.

Chemical modifications of lignin-based foams further enhance their sorption capacity. The addition of nanoparticles such as SiO_2 or Fe_3O_4 increases surface roughness and hydrophobicity, providing more adsorption sites and improving oil affinity. Wu et al. [11] demonstrated that lignin-based foam sorbents modified with SiO_2 nanoparticles exhibited increased oil adsorption capacity and surface hydrophobicity. Similarly, Shi et al. [19] showed that Fe_3O_4-modified lignin-based foam sorbents enhanced oil sorption capacity and introduced magnetic properties, facilitating easy recovery and reuse.

Lignin-based foam sorbents have shown exceptional performance in oil adsorption applications. Ma et al. [18] reported that carbon nanotube (CNT)-modified lignin-based foams could adsorb more than six times their weight in crude oil. The combination of lignin's inherent sorption properties and the photothermal effects of CNTs allowed the foam to operate efficiently in adsorption and recovery, as the CNTs helped reduce the oil's viscosity under solar irradiation.

The adaptability of lignin-based foam sorbents is also noteworthy. By adjusting the lignin content in the matrix, the foam's adsorption capacity can be tailored to target different contaminants. Higher lignin content increases the availability of adsorption sites and functional groups, enhancing the foam's ability to adsorb various organic pollutants [8].

The high sorption capacity of lignin-based foams is attributed to lignin's aromatic structure, functional groups, and the foam's increased porosity. Chemical modifications and the incorporation of additives further improve performance, making these foams suitable for a wide range of environmental applications. Their renewable and biodegradable nature also contributes to sustainability, offering an eco-friendly alternative to traditional sorbent materials.

4.2.4 PLA-Based Foam Sorbents

Polylactic acid (PLA)-based foam sorbents exhibit significant sorption capacities due to their highly porous structure, hydrophobicity, and mechanical properties. The open-cell architecture of these foams, often achieved through supercritical CO_2 foaming, creates a large surface area that can trap and absorb substantial amounts of oil. Li et al. [12] demonstrated that PLA–poly(butylene succinate) (PBS) composite foams fabricated using supercritical CO_2 foaming exhibited a porosity of 97.7% and an expansion ratio of 43.6. These structural characteristics contributed to high oil absorption capacities ranging from 7.9 to 21.9 g/g for various organic solvents and oils.

PLA's inherent hydrophobicity and oleophilic nature are pivotal in enhancing sorption capacity. PLA's hydrophobic nature allows it to repel water while attracting oil molecules, making it ideal for oil–water separation applications. Ding et al. [20] reported that adding modified lignin nanoparticles to PLA-based foams significantly enhanced their hydrophobicity, achieving a water contact angle of 141°. This strong hydrophobicity ensures that the foam remains unsaturated by water, maintaining its total absorption capacity for oils.

PLA's mechanical strength and structural rigidity also contribute to its performance as a sorbent material. The semicrystalline structure of PLA provides a balance between rigidity and flexibility, allowing the foam to retain its shape during absorption and desorption processes. Xiao et al. [21] explored the sorption capacity of PLA-based foams reinforced with Calotropis gigantea fiber, noting that the fiber-reinforced foams maintained structural integrity even after absorbing significant amounts of oil. This stability prevents foam collapse during use, enabling higher volumes of absorbed oil.

Thermal stability is another factor influencing the sorption capacity of PLA-based foam sorbents. Li et al. [12] showed that PLA–PBS foams exhibited excellent thermal stability up to 200 °C, allowing them to perform effectively in high-temperature or fluctuating environmental conditions. This resilience ensures that the foam maintains its porous structure and absorption capacity over multiple absorption–desorption cycles.

Surface chemistry modifications further enhance the sorption capacity. Functionalized nanoparticles, such as modified lignin, improve interaction with oil molecules by creating more active sites on the foam surface [20]. These modifications increase the foam's ability to capture and hold oil within its porous structure, improving overall sorption efficiency.

PLA's biodegradability adds an environmentally friendly dimension without compromising sorption capacity. Derived from renewable resources, PLA can biodegrade into non-toxic byproducts, making it suitable for environmentally sensitive areas like marine oil spill sites. Yang et al. [13] emphasized that the biodegradability of PLA-based foam sorbents does not negatively affect their oil absorption performance, with foams capable of absorbing up to 51.8 g/g of oil.

The sorption capacity of PLA-based foam sorbents is driven by their porous structure, hydrophobicity, mechanical strength, thermal stability, and surface chemistry. These factors make PLA-based foams highly effective in environmental remediation tasks, offering a reliable and sustainable solution for various sorption applications.

4.2.5 Biomass Oil-Based Foam Sorbents

Biomass oil-based foam sorbents, utilizing oils like castor and sunflower, exhibit high sorption capacities due to their porous structures, chemical compositions, and surface properties. Integrating natural oils into the matrix enhances environmental sustainability and oil absorption performance compared to conventional synthetic sorbents.

The porous structure resulting from using bio-oils as polyols creates an interconnected network of pores, allowing efficient absorption and retention of oils. Castor oil-based foam sorbents have an optimal pore size, facilitating rapid oil uptake while maintaining structural integrity [17]. This porosity, combined with hydrophobicity from oil-based components, enhances the foam's capacity to trap large quantities of oil.

The chemical nature of biomass oils contributes significantly. Oils like castor and sunflower contain hydroxyl groups participating in matrix formation, resulting in a structure that interacts favorably with oil molecules [17]. These hydroxyl groups strengthen the foam and oil interactions, improving capture and retention.

Incorporating lignin and other natural fillers enhances sorption capacity and foam stability. Castor oil–lignin-based foams benefit from lignin's rigid, aromatic structure, increasing their capacity to absorb various oil types. This combination creates a durable foam suitable for repeated industrial use.

Surface properties, particularly hydrophobicity, play a crucial role. Superhydrophobic foams based on castor oil demonstrate remarkable oil sorption capacities in oil–water separation. The hydrophobic surface repels water, preventing saturation and leaving more capacity for oil absorption, which is essential in marine oil spill applications.

Swelling behavior upon oil contact further enhances sorption capacity. Biomass oil-based foams are designed to swell, increasing oil absorption. Flexible polymer chains in the matrix expand as they interact with oil molecules [22]. Sunflower oil-based foams have demonstrated this capacity, absorbing their weight in oil several times.

The high sorption capacity of biomass oil-based foam sorbents results from their porous structures, chemical compositions, natural fillers, and hydrophobic surfaces. They offer durable and reusable solutions for oil spill remediation and industrial applications, combining effectiveness with environmental friendliness.

4.3 Hydrophobicity and Oleophilicity

Hydrophobicity and oleophilic are critical properties of bio-based sorbent foams, particularly in applications involving oil–water separation and environmental remediation. Hydrophobicity refers to the foam's ability to repel water, typically quantified by the water contact angle (WCA); a higher WCA indicates more excellent resistance to water infiltration. Oleophilicity describes the foam's affinity for oils, enhancing its capacity to absorb and retain oil substances. Together, these properties ensure selective oil absorption while repelling water, making these foams highly effective in aquatic environments where oil contamination is a concern.

The development of hydrophobic and oleophilic characteristics in bio-based foam sorbents depends on factors such as the chemical structure of the bio-based materials, surface modifications, incorporation of hydrophobic additives, and the foam's microstructure. The following sections explore how different bio-based foam sorbents exhibit enhanced hydrophobicity and oleophilicity due to their unique compositions and structural features. Table 4.1 summarizes the different water contact angles of bio-based foams.

4.3.1 Castor Oil-Based Foam Sorbents

Castor oil-based foam sorbents exhibit exceptional hydrophobicity and oleophilicity, making them highly effective for oil spill remediation and environmental cleanup. Castor oil, rich in ricinoleic acid, provides a unique combination of functional groups and hydrophobic chains that contribute to these properties.

The hydrophobicity of castor oil-based foams is attributed to the long hydrocarbon chains in ricinoleic acid, which reduce the foam's surface energy and repel water. Perera et al. [4] reported that modifying castor oil-based foams with octadecylsilane-modified diatomite particles enhanced the hydrophobicity, achieving water contact angles exceeding 150°, indicative of superhydrophobic surfaces. Incorporating these

Table 4.1 Water contact angles (WCA) of bio-based foam sorbents

Foam type	Water contact angle (°)	Hydrophobic modification
Coconut oil-based foam	139	None
Lignin-based foam with SiC nanoparticles	151.7	SiC nanoparticles
Fluorinated lignin-based foam	160.3	Fluorinated surface
Palm olein-based foam	145	None
Sunflower oil and sugarcane bagasse-based foam	138	None
Tannin-based foam	142	Natural tannins

hydrophobic particles increases surface roughness and creates hierarchical structures that trap air and minimize water contact.

Oleophilicity arises from the affinity between the foam's hydrophobic chains and oil molecules. The long fatty acid chains in castor oil interact favorably with hydrocarbons, allowing the foam to absorb oils effectively. Martins et al. [2] demonstrated that castor oil-based foams reinforced with palm fiber residues exhibited high oil sorption capacities, reaching up to 24 g/g for crude oil. The open-cell structure facilitates oil penetration and retention within the foam.

Incorporating silicon carbide (SiC) nanoparticles further enhances hydrophobicity and oleophilicity. Lv et al. [23] showed that castor oil and lignin-based foams modified with SiC nanoparticles achieved a water contact angle of 151.7°, indicating superhydrophobicity. The nanoparticles create a rough surface texture and contribute to the foam's thermal stability, making it suitable for use under varying environmental conditions.

Natural fillers like açaí seed biomass and coffee husks improve the foam's properties by increasing surface area and structural integrity. Maia et al. [3] found that adding coffee husk residues to castor oil-based foams enhanced oil sorption efficiency, exceeding 80% for diesel oils. The fillers contribute to the foam's oleophilicity and mechanical strength, allowing it to maintain performance over multiple absorption cycles.

The hydrolytic stability of castor oil-based foams ensures that the hydrophobic and oleophilic properties are retained even after prolonged exposure to water. The strong urethane bonds formed during polymerization resist degradation, making the foam durable for long-term applications.

Overall, castor oil-based foam sorbents achieve enhanced hydrophobicity and oleophilicity through the chemical structure of castor oil, surface modifications with hydrophobic particles, incorporation of nanoparticles, and addition of natural fillers. These properties enable the foams to effectively separate oils from water and maintain performance over repeated use, making them valuable materials for environmental remediation.

4.3.2 Other Plant Seed Oil-Based Foam Sorbents

Plant seed oil-based foam sorbents, synthesized from oils such as coconut, palm olein, and sunflower, exhibit excellent hydrophobicity and oleophilicity due to the bio-based polyols' fatty acid composition. The long-chain hydrocarbons in these oils contribute to forming hydrophobic surfaces that repel water and oleophilic surfaces that attract and absorb oils.

Coconut oil-based foam sorbents benefit from the high content of saturated fatty acids like lauric and myristic acids. These fatty acids have long hydrophobic alkyl chains that reduce the foam's surface energy, resulting in a water contact angle of approximately 139° [6]. The hydrophobicity ensures that water droplets form beads on the foam surface and roll off without wetting it.

The oleophilic nature of these foams is due to the compatibility between the hydrophobic alkyl chains and the hydrocarbon chains in oil molecules. This compatibility facilitates the absorption of oils into the foam matrix. The foam's open-cell structure, with interconnected pores, allows oil to penetrate and be retained within the foam. Tomon et al. [6] reported oil sorption capacities ranging from 14.89 to 24.65 g/g for coconut oil-based foams, demonstrating their effectiveness in oil absorption.

Palm olein-based foam sorbents also exhibit hydrophobic and oleophilic properties. Palm olein contains a mix of saturated and unsaturated fatty acids, such as palmitic and oleic. The presence of unsaturated bonds introduces slight variations in the hydrophobicity but maintains a low surface energy conducive to water repellency. Sittinun et al. [5] observed that palm olein-based foams had good hydrophobicity and were suitable for oil–water separation applications.

Sunflower oil-based foam sorbents further exemplify these properties. The polyols derived from sunflower oil possess hydroxyl groups that react with isocyanates to form urethane linkages, while the long hydrophobic chains contribute to surface hydrophobicity. Cuong et al. [7] incorporated sugarcane bagasse into sunflower oil-based foams, enhancing hydrophobicity and oleophilicity due to the synergistic effect of the hydrophobic matrix and the fibrous filler.

The surface morphology of plant seed oil-based foams plays a significant role in their hydrophobic and oleophilic behavior. The rough surface texture increases the surface area and creates hierarchical structures that enhance hydrophobicity through the lotus effect. Combining micro-scale roughness from the foam cells and nano-scale roughness from additives or fillers reduces the contact area between water droplets and the foam surface, promoting water repellency.

The foams' ability to maintain hydrophobicity and oleophilicity over multiple use cycles is essential for practical applications. The chemical stability of the fatty acid chains and the robustness of the urethane network ensure that the foam's surface properties are retained after repeated oil absorption and desorption processes. This durability contributes to the foams' cost-effectiveness and sustainability in long-term environmental remediation efforts.

4.3.3 Lignin-Based Foam Sorbent

Lignin-based foam sorbents exhibit remarkable hydrophobicity and oleophilic, essential for efficient oil–water separation and environmental remediation. Lignin, a complex aromatic polymer derived from plant biomass, contains a variety of functional groups such as hydroxyl (–OH), methoxy ($–OCH_3$), and phenolic structures. While lignin inherently possesses hydrophilic functional groups, its incorporation into the matrix and subsequent chemical modifications can significantly enhance the hydrophobic properties of the resulting foam.

One effective strategy to improve the hydrophobicity of lignin-based foam sorbents is the incorporation of hydrophobic nanoparticles. For instance, Wu et al. [11] modified lignin-based foam sorbents with SiO_2 nanoparticles through a simple dipping process. Adding these nanoparticles increased the surface roughness and decreased the surface energy of the foam, resulting in a superhydrophobic surface with a water contact angle of 151.3°. This superhydrophobicity is attributed to creating hierarchical micro- and nano-scale structures on the foam surface, which trap air and minimize the contact area between water droplets and the foam surface.

The foam's enhanced hydrophobicity prevents water from penetrating the porous structure, ensuring that the foam remains unsaturated by water and maintains its oil absorption capacity. This property is fundamental in oil spill remediation, where the foam must selectively absorb oil from an oil–water mixture without becoming waterlogged.

The oleophilicity of lignin-based foam sorbents is primarily due to the aromatic structure of lignin, which provides favorable interactions with non-polar oil molecules. The presence of aromatic rings and hydrophobic moieties in lignin facilitates the adsorption of oil substances. Additionally, the porous structure of the foam, resulting from the foaming process, offers a large surface area and numerous adsorption sites for oil molecules.

Functional modifications further enhance the foam's oleophilic properties. Ma et al. [9] demonstrated that lignin-based foams modified with carbon nanotubes (CNTs) exhibited improved oleophilic and hydrophobicity. The CNTs, with their high surface area and hydrophobic nature, synergistically interact with lignin to increase the foam's affinity for oils. Moreover, the CNTs impart photothermal properties to the foam, enabling solar-driven oil adsorption and facilitating oil recovery through photothermal heating.

The stability of the hydrophobic and oleophilic properties over multiple cycles is a critical aspect of practical applications. Lignin-based foam sorbents have shown excellent reusability due to the strong interactions between lignin, the matrix, and the hydrophobic additives. Shi et al. [19] reported that Fe_3O_4-modified lignin-based foam sorbents maintained their superhydrophobicity and oil absorption capacity after several adsorption–desorption cycles. The magnetic properties introduced by Fe_3O_4 nanoparticles also allowed for easy foam recovery using an external magnetic field.

The hydrophobicity and oleophilicity of lignin-based foam sorbents are achieved by combining lignin's aromatic structure, surface modifications with hydrophobic nanoparticles, and the foam's porous architecture. These properties enable the foam to selectively absorb oils while repelling water, making it highly effective for oil spill cleanup and wastewater treatment applications.

4.3.4 PLA-Based Foam Sorbents

Polylactic acid (PLA)-based foam sorbents exhibit significant hydrophobicity and oleophilicity, which is essential for their effectiveness in oil spill remediation and water separation. PLA is a biodegradable, aliphatic polyester derived from renewable resources like corn starch or sugarcane. It possesses inherent hydrophobic characteristics due to its non-polar methyl side groups.

The hydrophobicity of PLA-based foams can be enhanced through various methods, including surface modifications and the incorporation of hydrophobic additives. Ding et al. [20] improved the hydrophobicity of PLA foams by incorporating modified lignin nanoparticles. The lignin nanoparticles were grafted with long-chain alkyl groups to increase their hydrophobicity before blending into the PLA matrix. The resulting composite foam exhibited a water contact angle of 141°, indicating a highly hydrophobic surface [20].

Additionally, the foaming process itself influences the hydrophobicity. Creating a porous structure with microcellular or nanocellular pores increases surface roughness, which can enhance hydrophobicity due to the Cassie-Baxter effect. Li et al. [12] fabricated PLA-based foams using supercritical CO_2 foaming, resulting in open-cell structures with high porosity. The increased surface area and roughness contributed to the foam's water-repelling properties.

Oleophilicity in PLA-based foams arises from the polymer's affinity for non-polar substances. PLA's molecular structure allows favorable interactions with hydrocarbon chains present in oils. Incorporating additives such as Calotropis gigantea fiber (CGF) further enhances oleophilicity. Xiao et al. [21] reported that PLA foams reinforced with CGF displayed improved oil absorption capacities due to the fibrous structure's synergistic effect and the fibers' oleophilic nature.

Moreover, the porous architecture of PLA-based foams provides extensive surface area and interconnected pathways for oil infiltration. The open-cell structure facilitates rapid oil uptake and allows oil molecules to penetrate deep into the foam. This property is crucial for achieving high oil absorption capacities and efficient oil spill remediation.

The combination of hydrophobicity and oleophilicity ensures that PLA-based foams selectively absorb oils while minimizing water uptake. This selectivity is vital in oil–water separation applications where water can hinder the foam's performance. The hydrophobic surface prevents water from wetting the foam, maintaining its buoyancy and structural integrity during operation.

PLA's biodegradability adds environmental benefits to its application. Under appropriate conditions, PLA-based foams can degrade into non-toxic byproducts, reducing environmental pollution and contributing to sustainability. Yang et al. [13] highlighted that PLA-based foams retained their oil absorption performance while being biodegradable, making them suitable for applications in sensitive ecological areas.

4.3.5 Biomass-Based Foam Sorbents

Biomass-based foam sorbents utilizing oils such as castor and sunflower exhibit enhanced hydrophobicity and oleophilicity due to the presence of long hydrophobic alkyl chains and the foam's porous microstructure. The natural oils in these foams contain triglycerides with fatty acid chains that impart hydrophobic characteristics to the foam surface.

Castor oil-based foam sorbents benefit from ricinoleic acid, the primary component of castor oil, which contains hydroxyl groups and a long unsaturated hydrocarbon chain. The hydroxyl groups form the urethane network, while the hydrophobic chain contributes to water repellency. The oleophilic properties of biomass-based foams are enhanced by the compatibility between the hydrophobic chains in the bio-oils and the hydrocarbon chains in oils. The foam's surface readily interacts with oil molecules, facilitating absorption. The open-cell structure with interconnected pores allows oils to infiltrate and be retained within the foam matrix. Surface modifications can further improve hydrophobicity and oleophilicity. Incorporating hydrophobic nanoparticles or substances like lignin enhances surface roughness and reduces surface energy. Zhang et al. [16] showed that lignin-containing foam sorbents achieved superhydrophobic surfaces with water contact angles exceeding 150°, significantly improving oil–water separation efficiency.

The porous microstructure of biomass-based foams contributes to their hydrophobic and oleophilic properties. The foam cells create a rough surface that promotes hydrophobicity through the Cassie-Baxter effect. Additionally, the pores provide pathways for oil absorption, increasing the foam's capacity and rate of oil uptake.

The stability of these properties over multiple use cycles is essential for practical applications. Biomass-based foams have shown good reusability due to the urethane network's strong chemical bonds and the hydrophobic surface's stability. Using renewable and biodegradable materials also adds environmental benefits, aligning with sustainability goals.

4.4 Reusability and Mechanical Durability

Reusability, measured by the number of sorption–desorption cycles a foam can undergo without significant degradation, is crucial for sustainable applications in environmental remediation. High reusability minimizes waste and operational costs, making foam sorbents an economically viable option for long-term cleanup projects. Reusability is critical because it impacts the material's economic viability and environmental footprint. Materials that can be reused multiple times without significant degradation in performance reduce the frequency of replacement and disposal, lowering costs and minimizing waste. Table 4.2 summarizes the sorption capacities and number of cycles of different bio-based foams.

Table 4.2 Sorption capacity and reusability of bio-based foam sorbents

Foam type	Sorption capacity (g/g)	Reusability (cycles)
Coffee husk and castor oil foam	19.5	14
Macadamia waste and castor oil foam	7.3	30
Lignin-based foam with magnetic particles	27.2	100
Tannin-based foam	20.4	30
Sunflower oil and sugarcane bagasse-based foam	15.2	10
Palm olein-based foam	25.8	20
Coconut oil-based foam	24.5	20

4.4.1 Castor Oil-Based Foam Sorbents

The reusability of castor oil-based sorbent foams is a critical attribute that enhances their value for applications in oil spill cleanups and other environmental remediation efforts. Several factors contribute to their impressive reusability, including their structural integrity, surface properties, and the nature of the oils they absorb.

One main factor that makes castor oil-based sorbent foams highly reusable is their robust mechanical properties. These foams exhibit excellent elasticity and flexibility, which allow them to undergo multiple cycles of compression and expansion without significant structural degradation. For example, foam sorbents reinforced with coffee husk residues were shown to maintain their mechanical integrity over 11–14 cycles of oil absorption and desorption, with sorption efficiencies consistently exceeding 80% [3]. This mechanical resilience is crucial in maintaining the foam's capacity to absorb oils over repeated uses, ensuring it does not collapse or lose its porous structure.

The castor oil's chemical structure contributes to these foams' reusability. Castor oil contains a high proportion of ricinoleic acid, which provides multiple hydroxyl groups for forming urethane linkages with isocyanates. These urethane bonds contribute to the foam's stability and flexibility, allowing it to recover its original shape after compression [4]. This elasticity plays a crucial role in preventing permanent deformation of the foam, which is critical for maintaining pore accessibility and ensuring the foam can continue to absorb oils over multiple cycles.

Surface hydrophobicity is another critical factor that enhances the reusability of castor oil-based foams. The ability to repel water while selectively absorbing oil ensures that the foam can be used multiple times without becoming saturated with water, which would otherwise reduce its oil sorption capacity. Surface modifications, such as incorporating octadecylsilane-modified diatomite or SiC nanoparticles, significantly increase the hydrophobicity of the foam, as evidenced by contact angles exceeding 150° [23]. This superhydrophobicity allows the foam to separate oil from water over many cycles efficiently, preventing sorption efficiency degradation if water begins to infiltrate the foam's pores [4, 23].

Various fillers, such as palm residues, açaí seed biomass, and coffee husks, further enhance these foams' physical and chemical resilience. These agro-industrial residues not only improve the mechanical properties of the foam but also increase its oil sorption capacity and reusability. For example, foams containing palm fiber residues retained more than 50% of their sorption efficiency after 16 cycles of oil absorption and desorption for crude oil [2]. This is likely due to the improved pore morphology provided by the fillers, which increases the foam's ability to retain oil while maintaining its structural integrity during repeated use [2, 3].

Thermal stability also contributes to the reusability of castor oil-based foam sorbents. Thermal gravimetric analysis (TGA) has demonstrated that foams reinforced with nanoparticles, such as SiC, exhibit enhanced thermal stability, with decomposition temperatures higher than unmodified foams [23]. This thermal resilience allows the foams to be used in environments exposed to fluctuating temperatures or prolonged sun exposure, which could otherwise degrade the foam's material properties. The temperature of foams can reach up to 88.8 °C under sunlight exposure, aiding in the self-cleaning and rapid drying of the foams after oil absorption, thus extending their usable life [23].

In addition to their mechanical and thermal durability, the chemical properties of castor oil-based foam sorbents contribute to their reusability. These foams exhibit high resistance to chemical degradation, maintaining their sorption efficiency in a range of harsh environments, including acidic, saline, and alkaline conditions [24]. This chemical stability is essential for reusability in marine environments, where the foam may be exposed to seawater and other contaminants. For instance, foams with castor oil and agro-industrial residues like açaí seed biomass have shown sorption capacities exceeding 9 g/g for waste cooking oil, with good reusability over multiple cycles [2].

The combination of robust mechanical properties, enhanced hydrophobicity, thermal stability, and chemical resilience makes castor oil-based foam sorbents highly reusable. These foams can maintain their sorption efficiency over multiple oil absorption and desorption cycles, making them cost-effective and environmentally sustainable options for oil spill cleanups. Furthermore, incorporating agro-industrial residues enhances their performance and contributes to the circular economy by valorizing waste materials [3, 24].

Castor oil-based foam sorbents represent a significant advancement in the field of sustainable sorbent materials by reducing the need for frequent replacement and minimizing environmental impact. Future research should focus on further optimizing the formulation of these foams to maximize their reusability and exploring the potential of additional fillers and surface modifications to improve performance in specific environmental conditions.

4.4.2 *Plant Seed Oil-Based Foam Sorbents*

These foams, made from renewable resources like palm olein, coconut, and sunflower oil, offer a sustainable alternative to traditional petroleum-based foams. Their ability to maintain oil absorption capacity and structural integrity over multiple use cycles is driven by several inherent properties, including their hydrophobicity, mechanical strength, chemical stability, and porous structure.

A crucial factor that enhances the reusability of plant-seed-oil-based foam sorbents is their hydrophobicity and oleophilicity, which allows the foams to absorb oils while repelling water selectively. This is particularly important in environments where oil spills occur on water surfaces, as the foam must retain its oil sorption efficiency without becoming saturated with water. Coconut oil-based foam sorbents exhibit a water contact angle of approximately 139°, making them highly hydrophobic [6]. This water-repelling property ensures that the foams do not degrade due to waterlogging, enabling them to be reused multiple times for oil spill cleanups. The foam's ability to resist water absorption also prevents it from losing its structural integrity over repeated cycles of use.

The mechanical robustness of plant-seed-oil-based foam sorbents is another crucial factor contributing to their reusability. These foams exhibit good tensile and compressive strength, allowing them to endure the mechanical stresses associated with repetitive oil absorption and desorption. For example, coconut oil-based foam sorbents show a 47% increase in tensile strength compared to conventional foams, attributed to the high crosslinking density achieved through the plant-based polyols [6]. This enhanced mechanical strength is critical for ensuring the foam can be wrung or centrifuged to release absorbed oil without breaking down. Similarly, sunflower oil-based foam sorbents reinforced with sugarcane bagasse exhibit improved compressive strength, further supporting their durability in repeated-use applications [7].

The porous structure of plant-seed-oil-based foam sorbents also plays a pivotal role in their reusability. The open-cell structure of these foams provides a large surface area for oil absorption while allowing for easy release of the oil during recovery processes. Palm olein-based foam sorbents have an interconnected pore structure that enhances their oil sorption efficiency and ensures the foam maintains its performance even after multiple use cycles [5]. The porous architecture also helps distribute mechanical stress throughout the foam, preventing it from collapsing or losing its absorptive capacity after repeated oil absorption and desorption cycles.

Another critical property that enhances the reusability of plant-seed-oil-based foams is their chemical stability. These foams are designed to resist degradation when exposed to various chemicals, including oils and solvents. For example, palm olein-based foams retain their structural integrity and sorption performance even after exposure to engine oil and synthetic dyes like methyl orange [5]. This chemical resilience is crucial for maintaining the foam's reusability, as it ensures that the material does not degrade or lose efficiency when repeatedly exposed to different soil types and environmental conditions.

In addition to their structural and chemical properties, plant-seed-oil-based foam sorbents demonstrate excellent recyclability. Studies have shown that these foams can maintain their oil absorption capacities over multiple use cycles without significant performance loss. Coconut oil-based foam sorbents have been demonstrated to retain their sorption capacity after up to 20 cycles of oil absorption and desorption [6]. When reinforced with sugarcane bagasse, Sunflower oil-based foam sorbents also show high recyclability, with minimal loss in performance after repeated use [7]. This high level of recyclability makes plant-seed-oil-based foam sorbents a cost-effective and environmentally friendly alternative to conventional petroleum-based sorbents, which often degrade after only a few cycles of use.

The sustainability of plant-seed-oil-based foams further enhances their appeal for environmental applications. These foams are produced from renewable resources, reducing the reliance on petrochemical feedstocks and contributing to a lower carbon footprint. Their reusability aligns with sustainability principles, reducing the need for frequent replacement of sorbent materials, thereby minimizing waste generation.

The reusability of plant-seed-oil-based sorbent foams is driven by their hydrophobicity, mechanical strength, porous structure, chemical stability, and recyclability. These properties allow the foams to maintain their oil absorption efficiency and structural integrity over repeated use cycles, making them highly suitable for oil spill cleanup and wastewater treatment applications. The combination of performance and sustainability positions plant-seed-oil-based foam sorbents as an eco-friendly and durable solution to environmental challenges.

4.4.3 Lignin-Based Foam Sorbents

Lignin-based foam sorbents are promising candidates due to their inherent structural properties, mechanical resilience, and the incorporation of functional modifications that enhance their durability and efficiency over extended use. As a biopolymer, Lignin has a naturally complex, crosslinked aromatic structure, which imparts rigidity and strength to the foam. This structural integrity ensures the foam can withstand repeated physical stress, such as compression and deformation, during oil adsorption and recovery. Studies have shown that lignin-based foams maintain their shape and internal pore structure even after multiple cycles of compression, a critical factor in ensuring the foam continues to perform effectively [10]. Tests on modified lignin-based foams have demonstrated minimal mechanical strength and compressive resilience reduction, even after numerous compression cycles. This capacity to recover from compression without significant loss of structural integrity allows these foams to be reused multiple times without compromising their oil sorption efficiency [11].

The ability of lignin-based foam sorbents to retain their hydrophobic and oleophilic properties over repeated cycles is another critical factor contributing to their reusability. These surface characteristics are essential for selective oil adsorption in water–oil separation applications. The foam's ability to repel water while attracting and absorbing oil effectively removes contaminants from aqueous environments. Over time, repeated exposure to oil and water, combined with physical wear, could degrade these surface properties. However, lignin-based foam sorbents, particularly those modified with hydrophobic agents like SiO_2 nanoparticles, have demonstrated excellent stability in retaining their water-repellent properties [11]. Even after dozens of adsorption cycles, these foams maintain their superhydrophobicity, evidenced by a stable water contact angle above 150°. This durability ensures that the foam remains functional across multiple oil adsorption and desorption cycles.

Elastic recovery after compression is another crucial factor underpinning the reusability of lignin-based foam sorbents. In many applications, particularly oil spill cleanup, the foam is subjected to mechanical deformation as it absorbs oil and is later compressed to recover the absorbed oil. Relying on its original shape after compression allows the foam to maintain its porous structure, which is critical for repeated oil adsorption. Due to their inherent rigidity, lignin-based foam sorbents exhibit excellent elastic recovery, allowing them to withstand multiple compression and recovery cycles without collapsing or losing their functional porosity. This recovery capability ensures that the foams can be reused effectively, making them more sustainable and economically viable for long-term use [8].

The chemical stability of lignin-based foam sorbents also plays a significant role in their reusability. These foams are exposed to harsh chemical conditions in many industrial or environmental applications, such as alkaline or acidic environments. Lignin-based foams have demonstrated the ability to maintain their structural and functional integrity even after prolonged exposure to chemical agents, making them highly reusable in aggressive environmental conditions [19]. This chemical stability ensures that the foams do not lose effectiveness or degrade under chemical stress, enhancing their reusability in long-term applications.

Functional materials, such as carbon nanotubes (CNTs) and magnetic nanoparticles, also contribute to the enhanced reusability of lignin-based foam sorbents by improving their performance in specific applications. For example, CNT-modified lignin-based foam sorbents exhibit photothermal properties, allowing them to use sunlight to heat and reduce the viscosity of crude oil, facilitating faster and easier recovery of absorbed oil [18]. This ability to harness solar energy for oil recovery reduces the mechanical stress on the foam during the cleaning process and increases the foam's efficiency in repeated cycles. Similarly, magnetic nanoparticles such as Fe_3O_4 enable easy recovery using an external magnetic field. This functionalization simplifies the recovery process, ensuring that the foam can be cleaned and reused without significant wear or loss of performance [19].

While lignin-based foam sorbents exhibit impressive reusability, their biodegradability offers an added environmental benefit. Although these foams can be reused multiple times, their controlled biodegradability ensures environmentally responsible disposal at the end of their useful life. Unlike conventional petroleum-based foam sorbents, which persist in the environment for extended periods, lignin-based foams degrade relatively quickly under specific conditions, such as in alkaline environments. Fe_3O_4-modified lignin-based foam sorbents have been shown to degrade entirely in an alkaline methanol solution within 8 h [19]. This combination of high reusability and environmentally friendly disposal makes lignin-based foam sorbents attractive for sustainable environmental applications.

4.4.4 PLA-Based Foam Sorbents

One of the most critical aspects of the reusability of PLA-based sorbent foams is their mechanical durability. The PLA matrix provides a solid structural foundation, which allows the foam to endure the physical stresses encountered during repeated absorption and desorption cycles. For instance, Li et al. [12] noted that the open-cell PLA-PBS foams fabricated through supercritical CO2 foaming exhibited excellent mechanical strength. The structural integrity of these foams remained intact, even after multiple absorption cycles, which is essential for maintaining consistent oil absorption performance over time. The reinforcement of PLA with materials like poly(butylene succinate) (PBS) ensures that these foams do not degrade or lose their shape quickly during repeated use.

Hydrophobicity and oleophilicity are crucial characteristics that contribute significantly to the reusability of PLA-based foam sorbents. These foams demonstrate high hydrophobicity, which allows them to repel water while selectively absorbing oils, a vital trait in oil–water separation tasks. Hydrophobicity ensures the foam remains efficient by avoiding water saturation, which could reduce its oil absorption capacity. Ding et al. [20] highlighted that hydrophobicity could be further enhanced by incorporating additives such as modified lignin nanoparticles, which improve the foam's water-repelling ability. This property allows PLA-based foams to retain their oil absorption capacity even after multiple cycles, as water does not compromise the foams.

The porous structure of PLA-based foam sorbents is another crucial factor that enhances their reusability. A well-designed porous network provides a large surface area for oil absorption while facilitating the easy release of oil during desorption processes. Xiao et al. [21] demonstrated that PLA-CGF (Calotropis gigantea fiber) composite foams had a stable porous structure that absorbed large quantities of oil and retained its structural integrity after several cycles. The open-cell structure, in particular, as noted by Li et al. [12], allows for effective oil absorption and release, making the foam highly reusable. The interconnected pores provide pathways for oil to penetrate and be retained within the foam, while also ensuring that the foam can quickly release the absorbed oil during cleaning or reconditioning.

In addition to mechanical strength and porosity, thermal stability is another critical factor that supports the reusability of PLA-based sorbent foams. High thermal stability ensures that these foams can withstand variations in temperature without significant degradation, which is particularly important in applications where the foams may be exposed to heat during processing or cleaning. Li et al. [12] reported that PLA-PBS composite foams maintained their thermal stability up to 200 °C, indicating that they can be used in various environmental conditions without suffering performance loss. This thermal stability helps preserve the foam's structure and functionality during repeated absorption and desorption cycles.

Moreover, PLA-based foam sorbents' environmental sustainability enhances their reuse appeal. PLA is derived from renewable resources and is biodegradable, which means that even though these foams can be reused multiple times, they will eventually break down into non-toxic byproducts at the end of their lifecycle. This reduces the environmental impact compared to petroleum-based alternatives, making them a more sustainable option for long-term use. Yang et al. [13] emphasized the biodegradability of PLA-based foams, which ensures that these materials can be reused extensively without contributing to long-term environmental pollution.

4.4.5 Biomass Oil-Based Sorbent Foams

These foams, derived from renewable oils such as castor, Sunflower, and other plant-based oils, exhibit various properties that contribute to their durability and capacity for multiple uses. Their mechanical resilience, hydrophobicity, and structural stability make them well-suited for repeated absorption and regeneration cycles, reducing both environmental impact and operational costs.

A key feature of biomass oil-based foam sorbents is their mechanical durability, which enables them to undergo multiple compression and regeneration cycles without significant degradation. For example, castor oil-based foams have shown mechanical solid properties due to their high crosslink density and stable foam matrix, allowing them to be reused without losing structural integrity [17]. This mechanical resilience is crucial for maintaining the foam's oil absorption capacity across multiple uses, as it ensures that the foam can withstand the physical stresses of recovery processes, such as centrifugation or squeezing, without tearing or breaking apart.

In addition to mechanical strength, hydrophobicity is a central factor in the reusability of biomass-oil-based foam sorbents. These foams are designed to repel water while selectively absorbing oil, essential for oil spill remediation applications. The hydrophobic nature of these foams prevents water from saturating the material and diminishing its oil absorption efficiency. For instance, superhydrophobic foams based on castor oil and Lignin have demonstrated excellent oil absorption capacities while maintaining hydrophobicity over several reuse cycles. This ability to maintain hydrophobicity ensures that the foams can be regenerated effectively, allowing the absorbed oil to be easily removed without compromising their future performance.

Another aspect that enhances the reusability of biomass-oil-based foams is their capacity to retain absorbed oil during recovery. Castor oil-based foams have been shown to effectively trap oil within their porous structure, which prevents oil re-bleeding during handling or reuse. Integrating natural oils, which form a stable matrix within the foam structure, contributes to these materials' strong oil retention capabilities. This property is fundamental in applications where the foams are subjected to external pressure during recovery, as it ensures that minimal oil is lost, thus extending the foam's usability.

Thermal stability also plays a significant role in the reusability of biomass-oil-based foams. Many of these foams exhibit high thermal resistance, allowing them to be used in various environmental conditions without losing performance. Super-hydrophobic foams based on castor oil show remarkable thermal stability, which is critical for applications where the foams may be exposed to high temperatures during recovery or storage [17]. This thermal resilience enables the foams to be regenerated using heat or other thermal methods without degrading their structural properties, further extending their lifespan.

The ease of regeneration and the ability to reuse the foams multiple times is a significant advantage of biomass-oil-based sorbents. Castor oil-based foam sorbents have been demonstrated to retain their absorption efficiency after being regenerated up to 50 times using simple methods such as centrifugation. This high reusability makes these foams economically viable for large-scale applications, as they reduce the need for constant material replacement and minimize waste generation. Integrating bio-oils such as castor oil enhances the foam's performance and contributes to its biodegradability, making it an environmentally friendly option for long-term use.

Furthermore, the reusability of biomass-oil-based foam sorbents is supported by their ability to maintain consistent performance across different environmental conditions. Sunflower oil-based foam sorbents have been shown to perform well in batch and continuous-flow systems, maintaining their oil absorption efficiency over multiple cycles [22]. This versatility is essential for practical applications, where foams may be exposed to varying temperatures, pressures, and contaminants. By retaining their absorption capacity and mechanical strength in diverse conditions, these foams offer reliable performance over extended periods of use.

The reusability of biomass-oil-based sorbent foams can be attributed to their robust mechanical properties, high hydrophobicity, effective oil retention, and thermal stability. Combined with their eco-friendly nature and ease of regeneration, these characteristics make them ideal candidates for repeated use in environmental cleanup applications. Integrating bio-based oils such as castor, Sunflower, and other plant oils not only enhances the performance of these foams but also aligns with the growing demand for sustainable and renewable materials. As research continues, further improvements in foam design and material composition are expected to enhance their reusability, making biomass-oil-based foam sorbents a crucial tool in the fight against environmental pollution.

4.5 Thermal Stability

Thermal stability is a critical property of bio-based foam sorbents, mainly used in outdoor and industrial remediation settings where exposure to fluctuating or high temperatures is expected. The ability of these foams to maintain their structural

Table 4.3 Thermal stability of bio-based foams

Foam type	Decomposition temperature (°C)	UV stability
Coconut oil-based foam	350	Moderate
Castor oil with diatomite modification	340	High
Fluorinated lignin-based foam	355	High
Palm olein-based foam	330	Moderate
PLA-PBS foam with supercritical CO_2	320	Moderate
Honeycomb-like carbon foam	375	High

integrity and functional performance under thermal stress ensures their effectiveness and longevity in applications such as oil spill cleanup, industrial wastewater treatment, and insulation. Thermal stability is typically evaluated using techniques like thermogravimetric analysis (TGA) and differential scanning calorimetry (DSC), which measure the foam's resistance to thermal degradation and changes in physical properties at elevated temperatures. Understanding and enhancing the thermal stability of bio-based foam sorbents are essential for developing materials that can withstand harsh environmental conditions while maintaining sustainability and performance. Table 4.3 summarizes the thermal decomposition temperature and UV stability of bio-based foams.

4.5.1 Castor Oil-Based Foam Sorbents

Castor oil's unique chemical structure is a primary reason for these foams' good thermal stability. Castor oil contains a high percentage of ricinoleic acid, providing multiple hydroxyl groups for forming strong urethane bonds with isocyanates. These urethane linkages contribute to the robustness of the foam's molecular structure, enhancing resistance to thermal degradation [4]. The hydroxyl groups react with isocyanates to form stable cross-linked networks, helping the foam retain its shape and properties even under elevated temperatures.

Adding fillers to castor oil-based foam sorbents further enhances thermal stability. Incorporation of octadecylsilane-modified diatomite significantly improves resistance to thermal degradation. Castor oil-based foams grafted with this modified diatomite exhibit superior thermal stability, evidenced by higher decomposition temperatures in thermogravimetric analysis (TGA) compared to unmodified foams [4]. This enhancement is attributed to increased rigidity provided by the fillers, reinforcing the foam structure and slowing urethane bond breakdown under heat.

Nanoparticles like silicon carbide (SiC) also contribute to thermal stability. Including SiC nanoparticles in the foam matrix significantly increases thermal resistance. In a study focusing on castor oil and lignin-based foam sorbents with SiC nanoparticles, the foam exhibited excellent thermal stability, reaching temperatures as high as 88.8 °C under sunlight exposure [23]. This stability benefits applications that expose the foam to prolonged heat or sunlight, such as outdoor oil spill cleanups. Withstanding high temperatures without degradation ensures the foam remains effective longer in harsh environments.

Lignin, a natural polymer with high thermal stability, further enhances heat resistance. Lignin's three-dimensional network structure and functional groups contribute to its ability to withstand high temperatures, improving the thermal properties of foams in which it is incorporated [23]. Combining castor oil and lignin results in foams that possess good thermal stability and maintain mechanical properties under heat stress, making them suitable for environments with common temperature fluctuations.

Another factor contributing to thermal stability is resistance to oxidation at elevated temperatures. The cross-linked structure of the foams, combined with thermal stabilizers like nanoparticles, helps prevent the oxidation of polymer chains when exposed to heat. For example, foams reinforced with palm residues exhibit better thermal stability than unmodified foams, as fillers act as thermal barriers, slowing thermal degradation [2]. Enhanced resistance to oxidation ensures the foam does not degrade or lose oil sorption capacity when exposed to high temperatures during use.

Thermal stability is closely linked to the foam's ability to undergo multiple heating and cooling cycles without significant loss of structural integrity. Castor oil-based foam sorbents exhibit good thermal resilience, maintaining structural and functional properties over repeated thermal cycles. This is particularly important in applications where the foam may be subjected to varying temperatures, such as outdoor environments with daytime heating and nighttime cooling. Studies have shown that these foams can maintain mechanical properties and sorption efficiency over multiple thermal exposure cycles, highlighting their suitability for long-term use [23].

Furthermore, the ease of modifying castor oil-based foam sorbents with additives and fillers enhances heat resistance. Adding agro-industrial residues like açaí seed biomass and coffee husks improves thermal properties by increasing rigidity and reducing thermal degradation rates. These natural fillers act as reinforcing agents, providing additional stability to the foam structure at high temperatures [3]. Using renewable fillers improves thermal stability and enhances environmental sustainability, making them a greener alternative to conventional petroleum-based sorbents.

Castor oil-based foam sorbents' thermal stability results from their robust chemical structure, incorporation of thermal stabilizers and reinforcing fillers, and the natural properties of castor oil and lignin. These foams can withstand elevated temperatures without significant degradation, making them suitable for various applications, including oil spill cleanups in harsh environments. Maintaining structural integrity and sorption efficiency under heat stress ensures that castor oil-based foam sorbents remain effective over prolonged use, reducing the need for frequent replacement and lowering operational costs.

4.5.2 Plant Seed Oil-Based Foam Sorbents

Derived from renewable resources such as coconut oil, palm olein, and sunflower oil, these foams offer both sustainability and thermal resilience. Several factors contribute to their thermal stability, including the inherent properties of the bio-based polyols, cross-linking density, and the integration of natural reinforcements.

A primary contributor to the thermal stability of these foams is the molecular structure of the polyols used in their synthesis. Coconut oil-based polyols, for instance, contain predominantly saturated fatty acids, enhancing the foam's resistance to thermal degradation. Thermogravimetric analysis (TGA) has shown that coconut oil-based foam sorbents maintain structural integrity up to 351 °C, with initial degradation around 278 °C [6]. This high degradation temperature is attributed to the stable polyester-like structure of the coconut oil polyol, which is more thermally resilient than polyether-based polyols commonly used in petrochemical foams. The high hydroxyl number of coconut oil-based polyols facilitates better cross-linking during polymerization, further improving the foam's thermal stability.

Palm olein-based foam sorbents exhibit favorable thermal properties due to the composition of palm oil, which contains a mixture of saturated and unsaturated fatty acids. The polyols derived from palm olein slightly reduce the thermal degradation onset compared to petroleum-based foams but maintain sufficient stability for applications requiring moderate thermal resistance. Differential scanning calorimetry (DSC) studies indicate that palm olein-based foams have a glass transition temperature that supports their use in insulation and environmental remediation applications where thermal performance is critical [5]. The balance between flexibility and stability at elevated temperatures is achieved by combining unsaturated and saturated fatty acids in palm olein.

Sunflower oil-based foam sorbents exhibit good thermal stability, particularly when reinforced with natural materials like sugarcane bagasse. The bio-polyol derived from sunflower oil has a hydroxyl number of approximately 182 mg KOH/g, allowing effective cross-linking with isocyanates during foam synthesis [7]. This cross-linking contributes to the foam's thermal resilience. Adding sugarcane bagasse fibers enhances both mechanical properties and thermal stability. Lignocellulosic materials like bagasse provide additional reinforcement, allowing the foam to withstand higher temperatures before thermal degradation begins. This bio-composite foam resists thermal degradation up to approximately 320 °C, making it suitable for environments with moderate thermal stability.

Strong urethane linkages also contribute to the thermal stability of plant seed oil-based foam sorbents. The chemical structure of the foam consists of urethane bonds (–NH–CO–O–) formed through reactions between isocyanates and polyols. In these foams, the hydroxyl groups in the polyols, such as those from coconut and sunflower oils, promote the formation of solid urethane linkages. These thermally stable bonds enhance the foam's overall heat resistance. The high cross-linking density in coconut oil-based foams, resulting from the high hydroxyl number of the coconut oil polyol, significantly enhances thermal stability [6].

The foam's morphology also influences thermal stability. Plant seed oil-based foam sorbents' porous, interconnected cell structure facilitates oil absorption and thermal insulation properties. The open-cell structure allows heat dissipation throughout the foam, reducing the likelihood of localized thermal degradation. For example, palm olein-based foams exhibit a coarse cell structure with uniform pore distribution, helping mitigate heat exposure effects and maintain foam integrity at elevated temperatures [5].

Plant seed oil-based foam sorbents' thermal stability is influenced by the bio-based polyols' molecular structure, cross-linking density, and incorporation of natural reinforcements. Coconut oil-based foams exhibit the highest thermal resilience due to their saturated fatty acid content and high cross-link density. In contrast, sunflower oil-based foams reinforced with sugarcane bagasse offer thermal and mechanical enhancements. Palm olein-based foams provide sufficient stability for applications requiring moderate heat resistance. These characteristics make plant seed oil-based foam sorbents versatile and sustainable for applications requiring oil absorption and thermal performance.

4.5.3 Lignin-Based Foam Sorbents

Lignin is an aromatic polymer known for its high thermal stability, possessing a complex, cross-linked structure rich in phenolic hydroxyl groups that inherently resist heat-induced degradation. This property is a primary reason for the superior thermal performance of lignin-based foam sorbents. In foams, thermal stability is influenced by the polymer backbone and cross-linking density. The aromatic structure of lignin introduces additional cross-links into the matrix, resulting in a more thermally stable material. For instance, Mohammadpour and Sadeghi [8] demonstrated that incorporating higher concentrations of lignin polyol into foams increases the material's thermal resistance. Their thermogravimetric analysis (TGA) showed that lignin-based foams exhibit higher decomposition temperatures, especially with increased lignin content, due to lignin's ability to delay the degradation of the foam's molecular structure under heat [8].

The enhanced thermal properties of lignin-based foams are particularly evident when compared to foams synthesized with traditional petroleum-derived polyols. Wu et al. [11] demonstrated that lignin-based foams modified with SiO_2 nanoparticles exhibited superior oil adsorption properties and improved thermal stability. Incorporating lignin and nanoparticles created a highly cross-linked and reinforced foam structure, maintaining its integrity and functional properties under elevated temperatures [11]. This makes it ideal for oil–water separation applications in environments where temperature fluctuations are typical. The foam's ability to withstand higher temperatures without degradation allows it to be used in harsh industrial conditions where thermal resistance is essential for long-term functionality.

The integration of functional additives, such as carbon nanotubes (CNTs) and magnetic nanoparticles like Fe_3O_4, further enhances the thermal stability of lignin-based foam sorbents. For example, CNT-modified lignin-based foam sorbents developed by Ma et al. [18] exhibited remarkable thermal stability, partly attributed to CNTs' excellent heat dissipation properties. These foams, designed for solar-driven oil adsorption, could absorb and convert solar energy into heat without compromising structural integrity [18]. The CNTs facilitated even heat distribution throughout the foam, allowing it to withstand higher temperatures without degradation. This photothermal capability enabled the foam to maintain performance even under direct sunlight, where conventional foams would typically degrade more rapidly.

Similarly, Shi et al. [19] demonstrated that Fe_3O_4-modified lignin-based foam sorbents benefit from enhanced thermal stability due to the incorporation of magnetic nanoparticles. These nanoparticles impart magnetic properties and improve the foam's resistance to thermal degradation. The modified foams exhibited stable performance under high temperatures, making them suitable for industrial oil–water separation processes where thermal stability and durability are essential. Moreover, the magnetic properties facilitate the foams' recovery and reuse in high-temperature environments, contributing to their functionality and sustainability.

The thermal stability of lignin-based foam sorbents is also evident in their increased glass transition temperatures (Tg) with higher lignin content. Mohammadpour and Sadeghi [8] observed that the Tg of lignin-based foams rose significantly—from 38.5 to 90 °C—as more lignin was incorporated into the matrix [8]. This indicates that the foams can maintain their mechanical and structural properties at higher operating temperatures. The increase in Tg is directly related to the molecular rigidity and cross-linking introduced by lignin's aromatic rings, creating a more thermally stable foam network. The higher Tg suggests that lignin-based foams are well-suited for applications involving exposure to elevated temperatures, such as oil spill cleanups in tropical regions or industrial environments where heat is prevalent.

Additionally, lignin-based foams exhibit excellent thermal stability, withstanding long-term heat exposure without significant property changes. Wu et al. [11] reported that these foams maintained their superhydrophobicity and oil adsorption capacity even after exposure to high temperatures during oil–water separation processes. The thermal stability ensures that the foam can be reused multiple times under such conditions without heat-induced degradation. This durability enhances the foam's longevity and contributes to sustainability by reducing the need for frequent replacement.

The thermal stability of lignin-based foam sorbents is primarily driven by lignin's inherent aromatic structure, which enhances cross-linking and rigidity in the matrix. The integration of functional additives, such as SiO_2 nanoparticles, CNTs, and Fe_3O_4, further improves the foam's resistance to thermal degradation, making it suitable for high-temperature applications. These foams' increased glass transition temperatures and overall structural resilience ensure their performance in industrial and environmental settings where thermal stability is critical. The combination of lignin's natural properties and these enhancements make lignin-based foam sorbents superior in functionality and sustainability.

4.5.4 PLA-Based Foam Sorbents

One primary reason for the thermal stability of PLA-based foam sorbents is the inherent thermal resistance of PLA as a polymer. Derived from renewable resources like corn starch, PLA exhibits relatively high thermal stability compared to other biodegradable polymers. Its semi-crystalline nature provides a solid foundation for withstanding higher temperatures without significant degradation. Li et al. [12] reported that PLA–PBS composite foams fabricated through supercritical CO_2 foaming maintained their mechanical and absorption properties up to temperatures of 200 °C. This ability to resist thermal degradation is critical in applications where the foams are exposed to heat during oil absorption processes or in environments with fluctuating temperatures.

Adding poly(butylene succinate) (PBS) to PLA matrices further enhances the thermal stability of the foams. PBS, a biodegradable polymer with good flexibility, not only improves the mechanical properties of PLA but also contributes to stabilizing its thermal behavior. The combination of PLA and PBS results in foams that can endure the stresses associated with high temperatures while maintaining their porous structure and oil absorption capacity. Li et al. [12] showed that this blend produced foams with superior thermal resilience compared to re-PLA, allowing them to function effectively in harsher environmental conditions.

Structural modifications, such as fiber reinforcement, also contribute to the thermal stability of PLA-based foam sorbents. Xiao et al. [21] highlighted that PLA-based foams reinforced with *Calotropis gigantea* fibers exhibited stable thermal properties due to the structural integrity imparted by the fibers. The fiber reinforcement improved the foam's mechanical performance and helped it retain thermal properties under stress. The porous structure created during the foaming process facilitates heat dissipation, preventing localized heating and ensuring the foam does not collapse or lose structural integrity when exposed to elevated temperatures during oil absorption processes.

The incorporation of additives into the PLA matrix can further improve thermal stability. Ding et al. [20] demonstrated that adding modified lignin nanoparticles significantly enhanced the thermal properties of PLA-based foam sorbents. Lignin, a natural polymer with high thermal resistance, provided additional protection against thermal degradation. The modified lignin nanoparticles reinforced the foam matrix, improving its resistance to heat-induced deformation. The thermal degradation temperature of the PLA–lignin composite foams was notably higher than that of pure PLA foams, showcasing lignin's potential as a thermal stabilizer.

Advanced fabrication techniques, such as supercritical CO_2 foaming, also significantly determine the thermal stability of PLA-based foam sorbents. This solvent-free process produces highly porous foams with improved thermal properties. The method allows precise control over the foam's cell structure, resulting in a material that can better withstand thermal fluctuations. The interconnected porous structure increases the foam's surface area, facilitating heat dissipation and preventing localized overheating. This improves the foam's absorption capacity and ensures it retains thermal stability during extended use.

The crystallinity of PLA contributes to the thermal stability of the foam. Higher degrees of crystallinity generally lead to improved thermal resistance. Yang et al. [20] explored the role of poly(3-hydroxybutyrate-co-3-hydroxyvalerate) (PHBV) in enhancing the crystallization of PLA-based foam sorbents. PHBV facilitated the formation of a more stable crystalline structure within the PLA matrix, improving the foam's thermal stability. The increased crystallinity allowed the foam to maintain its structure and oil absorption capabilities even when exposed to elevated temperatures, further contributing to its reusability and long-term performance.

The thermal stability of PLA-based foam sorbents results from the inherent properties of PLA, structural modifications such as fiber reinforcement, the addition of thermal stabilizing additives like lignin, and advanced fabrication techniques like supercritical CO_2 foaming. These factors enable the foams to withstand higher temperatures without significant degradation, making them highly effective in applications requiring thermal resilience. The ability to maintain structural and functional integrity at elevated temperatures ensures that these foams are not only reusable but also reliable in demanding environmental conditions.

4.5.5 *Biomass-Based Foam Sorbents*

The thermal stability of biomass-based sorbent foams significantly contributes to their versatility and performance in environmental remediation and industrial applications. Biomass-based foam sorbents, utilizing oils like castor and sunflower, exhibit enhanced thermal stability due to the inherent properties of bio-oils, which contribute to the development of cross-linked polymer networks resistant to heat-induced degradation. This thermal resilience makes these foams particularly useful in environments with temperature fluctuations or where thermal processing is required for foam regeneration.

One primary reason for biomass-based foam sorbents' improved thermal stability is the chemical composition of the bio-oils used as polyol components. Bio-oils like castor oil contain hydroxyl groups that participate in forming highly cross-linked polymer structures during synthesis. Introducing such rigid cross-linked networks results in higher thermal degradation resistance than conventional petroleum-based. The solid intermolecular forces within the cross-linked network limit polymer chain mobility, requiring more energy to break bonds, directly contributing to higher thermal stability.

Incorporating natural additives and nanoparticles further enhances the thermal stability of biomass-based foam sorbents. For instance, foams that combine castor oil with lignin or other natural fillers exhibit increased thermal decomposition temperatures due to the synergistic effect of lignin's natural heat resistance [17]. Lignin, an aromatic polymer, contributes to the foam's structural integrity under elevated temperatures, slowing the onset of thermal degradation. Combining lignin and biomass oils creates a composite material with enhanced thermal properties suitable for high-temperature applications.

Maintaining hydrophobicity under thermal stress also contributes to the thermal stability of biomass-based foam sorbents. Hydrophobicity prevents moisture absorption, which can lead to premature thermal degradation. Superhydrophobic foams based on castor oil have demonstrated excellent thermal stability, maintaining hydrophobic surface characteristics even when exposed to high temperatures. This feature is beneficial in environments where heat and moisture are present, ensuring the foam's thermal performance is not compromised by water absorption.

Additionally, tannin-based and sunflower oil-based foam sorbents exhibit significant thermal stability due to the bio-oil's natural antioxidant properties, inhibiting oxidative degradation at high temperatures [22]. These foams display a delayed onset of thermal decomposition, with degradation temperatures significantly higher than conventional foams. This delay is essential in industrial settings where foams are exposed to fluctuating temperatures, allowing effective performance over extended periods without breakdown.

The pyrolytic behavior of biomass-based foam sorbents also contributes to their thermal resilience. During thermal decomposition, foams derived from biomass oils tend to form a protective char layer, acting as an insulating barrier against further degradation. For example, liquefied-larch-based carbon foams produce a stable carbonized structure resisting further thermal breakdown, providing excellent thermal stability [14]. The formation of this carbonaceous char layer slows thermal decomposition, allowing the foam to maintain structural integrity under high-temperature conditions.

The thermal stability of biomass-based sorbent foams can be attributed to cross-linked polymer networks formed by bio-oils, incorporating natural additives like lignin, maintaining hydrophobicity, and pyrolytic behavior leading to char formation. Under elevated temperatures, these foams resist thermal degradation while retaining functional properties, such as oil absorption and structural integrity. As a result, they offer robust and sustainable solutions for applications in oil spill remediation and industrial processes where thermal stability is crucial.

4.6 Sorption Isotherms

Sorption isotherms are fundamental tools in characterizing the interactions between sorbents and sorbates, providing vital quantitative insights into adsorption mechanisms, capacities, and the surface properties of materials. Understanding these isotherms is crucial for optimizing foam properties to enhance oil absorption efficiency, particularly in environmental remediation applications. The ability to predict how these foams will interact with various oil contaminants under different conditions allows researchers and engineers to design materials that are both effective and sustainable.

The adsorption process in sorbent materials is often complex, involving various physical and chemical interactions between the sorbent and sorbate molecules. To model and understand these interactions, several sorption isotherm models are employed. The most commonly used models are the Langmuir, Freundlich, and Redlich–Peterson isotherms, each offering different perspectives on the adsorption process based on specific assumptions about the nature of the sorbent surface and the adsorption mechanism.

The Langmuir isotherm model assumes monolayer adsorption onto a homogeneous surface with a finite number of identical sites. It posits that once a sorbate molecule occupies a site, no further adsorption can occur at that site, implying uniform energies of adsorption without interaction between adsorbed molecules. The Langmuir equation is expressed as:

$$q_e = \frac{q_{\max} K_L C_e}{1 + K_L C_e}$$

where:

q_e is the amount of adsorbate adsorbed per unit mass of adsorbent at equilibrium (mg/g),
$q_{\max}$ represents the maximum adsorption capacity (mg/g),
K_L is the Langmuir constant related to the affinity of binding sites (L/mg),
C_e is the equilibrium concentration of the adsorbate in solution (mg/L).

This model is particularly useful when the adsorption occurs on a structurally uniform surface and is limited to one molecular layer.

The Freundlich isotherm model is an empirical equation that assumes adsorption occurs on a heterogeneous surface with sites that have different energies of adsorption, leading to multilayer adsorption. It is suitable for modeling adsorption on surfaces with a non-uniform distribution of heat of adsorption and affinities over the heterogeneous surface. The Freundlich equation is given by:

$$q_e = K_F C_e^{\frac{1}{n}}$$

where:

K_F is the Freundlich constant indicative of adsorption capacity [(mg/g)(L/mg)^(1/n)],
n is the heterogeneity factor indicating the favorability and intensity of the adsorption process.

A smaller n value (closer to zero) indicates a more heterogeneous surface and a stronger interaction between the sorbent and sorbate.

The Redlich–Peterson isotherm model is a hybrid that incorporates features of both the Langmuir and Freundlich isotherms, accommodating adsorption systems that do not strictly follow either model. It is suitable over a wide concentration range and can describe adsorption on both homogeneous and heterogeneous surfaces. The Redlich–Peterson equation is:

$$q_e = \frac{K_R C_e}{1 + \alpha_R C_e^{\beta}}$$

where:

K_R (L/g) and α_R (L/mg)	are Redlich–Peterson constants,
β	is an exponent between 0 and 1.

This model is particularly adaptable, making it applicable to systems where the adsorption mechanism is complex and cannot be adequately described by the Langmuir or Freundlich models alone.

Understanding and applying these models is fundamental for interpreting adsorption data and optimizing the design and functionality of bio-based foams. By fitting experimental data to these isotherm equations, researchers can determine key parameters such as the maximum adsorption capacity and the affinity constants, which provide insights into the efficiency and effectiveness of the sorbent material.

The choice of an appropriate isotherm model provides deeper understanding into the adsorption mechanism—whether it involves monolayer or multilayer adsorption, homogeneous or heterogeneous surface energies, and physisorption or chemisorption processes. This knowledge is critical for tailoring foam characteristics to meet specific industrial and environmental needs, such as enhancing oil absorption capacity, improving selectivity for certain contaminants, or increasing the durability and reusability of the sorbent material.

In the following sections, we delve into the sorption isotherms of various bio-based foams, exploring how their unique compositions and structural characteristics influence their adsorption behaviors and capacities. We examine lignin-based, PLA-based, plant seed oil-based, biomass-based, and castor oil-based foams, analyzing how sorption isotherm models apply to each and how this knowledge aids in optimizing their performance for environmental applications such as oil spill remediation and wastewater treatment.

4.6.1 Castor Oil-Based Foam Sorbents

The Langmuir isotherm often aptly describes these foams, especially when surface modifications enhance sorption capacities. Foams reinforced with fillers like palm fiber residues fit well with the Langmuir model, indicating monolayer adsorption and uniform surface sites [2]. The incorporation of fillers improves mechanical properties and creates more uniform pores, facilitating consistent adsorption.

In cases where the foam surface is heterogeneous—due to fillers like coffee husk residues—the Freundlich isotherm becomes more applicable, reflecting multilayer adsorption on surfaces with varied energies [3]. The irregularities introduced by the fillers create a range of adsorption sites, leading to a nonlinear increase in sorption capacity with concentration.

Surface modifications, such as incorporating hydrophobic nanoparticles, can shift the sorption behavior towards the Langmuir model by creating more uniform adsorption sites [4]. Enhancing surface hydrophobicity increases the affinity for oil molecules, and the uniform distribution of nanoparticles leads to consistent adsorption energies across the surface.

Understanding the appropriate sorption isotherm model is crucial for optimizing foam design and predicting performance under different oil concentrations. It informs the maximum capacity of the foams and guides the selection of fillers and surface treatments to enhance efficiency in environmental cleanup operations.

Moreover, thermodynamic parameters derived from isotherm models, such as the maximum adsorption capacity (qmax) and adsorption constants, provide valuable information for scaling up the application of these foams. High values indicate a foam's suitability for absorbing large volumes of oil, essential for addressing significant spill events.

The reusability of castor oil-based foams, supported by stable sorption isotherms over multiple cycles, highlights their economic and environmental benefits. Consistent performance reduces the need for frequent replacement, lowering operational costs and minimizing waste generation.

4.6.2 Plant Seed Oil-Based Foam Sorbents

Plant seed oil-based foam sorbents, derived from renewable sources like coconut, palm olein, and sunflower oils, display sorption behaviors influenced by their chemical composition, pore structure, and surface properties. Their inherent hydrophobicity and oleophilicity result from the bio-based polyols that promote strong oil affinity. The fatty acid chains present in these oils impart hydrophobic characteristics, enhancing the foam's ability to absorb oils over water selectively.

Coconut oil-based foams, for example, exhibit high sorption capacities due to their superoleophilic surfaces and porous structures, with oil absorption capacities ranging from 14.89 to 24.65 g/g depending on the oil type [6]. The long-chain fatty acids in coconut oil contribute to the low surface energy of the foam, facilitating oil spread and absorption.

The sorption isotherms of these foams often fit the Freundlich model, indicating multilayer adsorption on heterogeneous surfaces. This is attributed to the complex surface structures created by the combination of bio-polyols and fillers like sugarcane bagasse, which introduce varied active sites with different affinities for oil molecules [7]. The presence of fillers not only enhances mechanical properties but also increases surface roughness and porosity, providing additional adsorption sites.

In cases where the foam exhibits more uniform surface characteristics, the Langmuir isotherm may be applicable. For palm olein-based foams with relatively homogeneous pore structures, the Langmuir model accurately describes the monolayer adsorption process [5]. This suggests that optimizing the foam's synthesis parameters to control pore size and distribution can tailor the adsorption behavior to specific needs.

The presence of hydroxyl groups in the polyols enhances urethane linkages during synthesis, providing active sites that increase sorption capacity and influence the isotherm profile. The high hydroxyl value of sunflower oil-based polyols contributes to stronger interactions with oil molecules, as these groups can engage in hydrogen bonding and other intermolecular forces, enhancing adsorption efficiency [7].

The recyclability of plant seed oil-based foams contributes to consistent sorption behavior over multiple use cycles. Their ability to retain sorption capacities after repeated absorption and desorption ensures predictable performance in practical applications, such as oil spill cleanup. The stability of the sorption isotherms over multiple cycles indicates that the foam's structural integrity and surface properties remain intact, which is crucial for long-term environmental remediation efforts.

4.6.3 Lignin-Based Foam Sorbents

Lignin-based foams exhibit complex adsorption behaviors due to their heterogeneous and porous structures, which provide extensive surface areas for oil adsorption. The inherent rigidity and aromatic nature of lignin contribute to the foam's structural integrity and adsorption capabilities. The Langmuir isotherm, which assumes monolayer adsorption onto a surface with a finite number of identical sites, has been effectively applied to these foams. For instance, Mohammadpour and Sadeghi [8] found that the Langmuir model accurately described the oil sorption behavior of lignin-based foams, with maximum adsorption capacity increasing alongside lignin content. This suggests efficient utilization of adsorption sites attributed to lignin's rigid structure and uniform distribution of active sites [8].

However, when surface modifications introduce heterogeneity—such as the addition of SiO_2 nanoparticles—the Freundlich isotherm becomes more appropriate. Wu et al. [11] demonstrated that the Freundlich model better fit the sorption data for lignin-based foams modified with SiO_2, indicating multilayer adsorption on heterogeneous surfaces. The incorporation of nanoparticles creates a more irregular surface topology, introducing a variety of adsorption sites with different energies, which the Freundlich model effectively captures.

The Redlich–Peterson isotherm, a hybrid of the Langmuir and Freundlich models, has also been employed for lignin-based foams modified with carbon nanotubes (CNTs). Ma et al. [18] found that this model effectively described the dual characteristics of monolayer and multilayer adsorption exhibited by CNT-modified foams. The uniform dispersion of CNTs contributes to homogeneous adsorption sites, while the intrinsic porosity of lignin introduces heterogeneity. The Redlich–Peterson model accommodates this complexity by accounting for both homogeneous and heterogeneous adsorption processes.

These isotherm analyses reveal that lignin-based foam sorbents engage in both physisorption and chemisorption. The presence of functional groups such as hydroxyl and carboxyl groups in lignin facilitates chemical interactions with oil molecules, enhancing sorption capacity through chemisorption. Physisorption occurs via van der Waals forces, allowing oils to adhere physically to the foam's surface. Functional additives like nanoparticles further expand available adsorption sites, improving efficiency across various applications [19]. Sorption isotherms also highlight the influence of surface hydrophobicity and pore structure on adsorption efficiency, with larger surface areas and pore volumes correlating with increased oil adsorption [11].

Moreover, the thermodynamic parameters derived from these isotherms, such as the Langmuir constant K_L and Freundlich constants K_F and n, provide insights into the affinity and capacity of the foam for oil molecules. Higher values of K_L and K_F indicate stronger interactions and higher sorption capacities, guiding the selection of appropriate lignin content and surface modifications to enhance performance.

4.6.4 PLA-Based Foam Sorbents

For polylactic acid (PLA)-based foam sorbent, the sorption isotherm is significantly influenced by the foam's porous structure, surface chemistry, and polymer–oil interactions. The open-cell architecture achieved through techniques like supercritical CO_2 foaming results in high porosity and interconnected pores, enhancing oil sorption capacity [12]. The foam's porosity provides numerous adsorption sites and facilitates the penetration of oil molecules into the foam matrix, leading to higher absorption rates and capacities.

The hydrophobic and oleophilic nature of PLA, often augmented with additives like lignin nanoparticles, facilitates selective oil absorption while repelling water [20]. This selective affinity is critical in oil–water separation applications, where the foam must efficiently absorb oil without becoming saturated with water. Modifying PLA with hydrophobic additives increases surface roughness and decreases surface energy, enhancing oil affinity.

The interaction between the PLA matrix and absorbed oils is critical. Reinforcements such as *Calotropis gigantea* fiber (CGF) improve mechanical properties and provide additional binding sites, contributing to efficient sorption processes [21]. The fibers introduce micro- and nanoscale roughness on the foam surface, increasing the contact area for oil adsorption. Additionally, the compatibility between the PLA matrix and CGF ensures that the structural integrity of the foam is maintained during absorption, preventing collapse or degradation.

Thermodynamic factors, including the Gibbs free energy of adsorption, affect sorption kinetics and isotherm profiles. Modifications that increase the crystallinity of PLA, such as incorporating poly(3-hydroxybutyrate-co-3-hydroxyvalerate) (PHBV), enhance the foam's capacity to maintain equilibrium at higher oil concentrations [13]. Higher crystallinity leads to a more rigid and stable foam structure, resisting deformation during oil uptake and improving reusability.

Empirical data often align with the Freundlich isotherm, suggesting multilayer adsorption on heterogeneous surfaces due to the foam's high porosity and variable surface sites [12]. The heterogeneous nature arises from the distribution of pore sizes and the presence of additives, creating a range of adsorption energies. The Freundlich model accommodates the non-uniformity of the adsorption process, which is characteristic of PLA-based foams.

Furthermore, the sorption capacity and efficiency are influenced by the oil's viscosity and surface tension. Low-viscosity oils penetrate the foam matrix more readily, resulting in higher sorption capacities. Understanding these interactions allows for the optimization of PLA-based foams for specific oil types encountered in environmental spills.

4.6.5 Biomass-Based Foam Sorbents

Biomass oil-based foam sorbents, utilizing oils like castor and sunflower, exhibit sorption isotherms that shed light on their absorption mechanisms and efficiency. The unique chemical structures of these oils, rich in hydroxyl groups and fatty acid chains, influence the foam's interaction with oil contaminants.

The Langmuir isotherm frequently applies, suggesting monolayer adsorption onto surfaces with finite identical sites. This model has been effective in describing the sorption behavior of foams synthesized from castor oil, indicating specific active sites for oil interaction. The uniformity of the foam's surface, resulting from the consistent chemical composition of castor oil, supports the monolayer adsorption assumption.

The Freundlich isotherm also plays a role, particularly for foams with heterogeneous surfaces due to natural oil components introducing structural variability. This model accounts for multilayer adsorption and varying sorption sites, aligning with the characteristics of biomass oil-based foams [22]. The heterogeneity arises from variations in polymerization and crosslinking during foam synthesis, leading to a distribution of pore sizes and surface energies.

The Redlich–Peterson isotherm combines both models, offering a versatile description of sorption across a range of concentrations and has been successfully applied to these foams. This model's adaptability makes it suitable for systems where neither the Langmuir nor the Freundlich isotherm alone provides an adequate fit, capturing the complexity of adsorption processes in biomass-based foams.

Sorption isotherms reveal that the chemical nature of biomass oils enhances interactions with oil contaminants, improving sorption performance. The presence of ricinoleic acid in castor oil introduces hydroxyl functionalities that can form hydrogen bonds with oil molecules, increasing adsorption affinity [4]. Additionally, the flexible aliphatic chains in these oils contribute to the foam's elasticity and ability to conform to oil droplets, enhancing contact and absorption.

The maintenance of stable sorption properties over multiple regeneration cycles further emphasizes the suitability of these foams for sustainable environmental applications. The robustness of the foam structure ensures that the active sites remain available after desorption processes, which often involve mechanical squeezing or solvent washing. This durability is reflected in consistent sorption isotherms across cycles, indicating reliable performance.

References

1. Martins, L.S., Silva, N.G.S., Claro, A.M., Amaral, N.C., Barud, H.S., Mulinari, D.R.: Insight on açaí seed biomass economy and waste cooking oil: eco-sorbent castor oil-based. J. Environ. Manage. **293**, 112803 (2021). https://doi.org/10.1016/j.jenvman.2021.112803
2. Martins, L.S., et al.: Crude oil and S500 diesel removal from seawater by polyurethane composites reinforced with palm fiber residues. Chemosphere **267**, 129288 (2021). https://doi.org/10.1016/j.chemosphere.2020.129288
3. Maia, L.S., et al.: Coffee husks residues incorporated into polyurethane foam towards greener material for diesel S10 and S500 removal and recovery. Ind. Crops Prod. **189**, 115747 (2022). https://doi.org/10.1016/j.indcrop.2022.115747
4. Perera, H.J., Goyal, A., Alhassan, S.M., Banu, H.: Biobased castor oil-based polyurethane foams grafted with octadecylsilane-modified diatomite for use as eco-friendly and low-cost sorbents for crude oil clean-up applications. Polymers **14**(23), 5310 (2022). https://doi.org/10.3390/polym14235310
5. Sittinun, A., Pisitsak, P., Manuspiya, H., Thiangtham, S., Chang, Y.-H., Ummartyotin, S.: Utilization of palm olein-based polyol for polyurethane foam sponge synthesis: potential as a sorbent material. J. Polym. Environ. **28**(12), 3181–3191 (2020). https://doi.org/10.1007/s10924-020-01834-4
6. Tomon, T.R.B., et al.: A novel naturally superoleophilic coconut oil-based foam with inherent hydrophobic properties for oil and grease sorption. Sci. Rep. **14**(1), 14223 (2024). https://doi.org/10.1038/s41598-024-64178-2
7. Cuong, T.D., Dien, L.Q., Hoang, P.H.: Preparation of bio-based porous material with high oil adsorption capacity from bio-polyurethane and sugarcane bagasse. RSC Adv. **14**(10), 6938–6947 (2024). https://doi.org/10.1039/D4RA00469H
8. Mohammadpour, R., Mir Mohamad Sadeghi, G.: Effect of liquefied lignin content on synthesis of bio-based polyurethane foam for oil adsorption application. J. Polym. Environ. **28**(3), 892–905 (2020). https://doi.org/10.1007/s10924-019-01650-5
9. Ma, X., et al.: Mechanically robust, solar-driven, and degradable lignin-based polyurethane adsorbent for efficient crude oil spill remediation. Chem. Eng. J. **415**, 128956 (2021). https://doi.org/10.1016/j.cej.2021.128956
10. Chen, J., et al.: Multifunctional superhydrophilic/underwater superoleophobic lignin-based polyurethane foam for highly efficient oil-water separation and water purification. Sep. Purif. Technol. **311**, 123284 (2023). https://doi.org/10.1016/j.seppur.2023.123284
11. Wu, J., et al.: Superhydrophobic lignin-based multifunctional polyurethane foam with SiO_2 nanoparticles for efficient oil adsorption and separation. Sci. Total. Environ. **860**, 160276 (2023). https://doi.org/10.1016/j.scitotenv.2022.160276
12. Li, B., Zhao, G., Wang, G., Zhang, L., Gong, J., Shi, Z.: Biodegradable PLA/PBS open-cell foam fabricated by supercritical CO_2 foaming for selective oil-adsorption. Sep. Purif. Technol. **257**, 117949 (2021). https://doi.org/10.1016/j.seppur.2020.117949
13. Yang, H., et al.: Fabrication of bio-based biodegradable poly(lactic acid) (PLA) and poly(3-hydroxybutyrate-co-3-hydroxyvalerate) (PHBV) composite foams for highly efficient oil-water

separation. Int. J. Biol. Macromol. **257**, 128750 (2024). https://doi.org/10.1016/j.ijbiomac.2023.128750

14. Tan, J., Li, W., Ma, C., Wu, Q., Xu, Z., Liu, S.: Synthesis of honeycomb-like carbon foam from larch sawdust as efficient absorbents for oil spills cleanup and recovery. Materials **11**(7), 1106 (2018). https://doi.org/10.3390/ma11071106
15. Songsaeng, S., Thamyongkit, P., Poompradub, S.: Natural rubber/reduced-graphene oxide composite materials: morphological and oil adsorption properties for treatment of oil spills. J. Adv. Res. **20**, 79–89 (2019). https://doi.org/10.1016/j.jare.2019.05.007
16. Zhang, T., Yu, M., Huang, Y., Tan, J., Zhang, M., Zhu, X.: Design and manufacturing of cost-effective tannin-based polyurethane foam as an efficient and reusable absorbent for oil and solvents. Ind. Crops Prod. **189**, 115815 (2022). https://doi.org/10.1016/j.indcrop.2022.115815
17. Wang, Y., Yu, X., Fan, W., Liu, R., Liu, Y.: Alginate-oil gelator composite foam for effective oil spill treatment. Carbohydr. Polym. **294**, 119755 (2022). https://doi.org/10.1016/j.carbpol.2022.119755
18. Ma, S., et al.: Multifunctional hydrophobic bio-based foam toward highly efficient photothermal cleanup of high-viscosity crude oil spills, oil-water separation and aqueous organic pollutant elimination. Colloids Surf. Physicochem. Eng. Asp. **667**, 131431 (2023). https://doi.org/10.1016/j.colsurfa.2023.131431
19. Shi, Y., et al.: One-step foam assisted synthesis of robust superhydrophobic lignin-based magnetic foams for highly efficient oil-water separation and rapid solar-driven heavy oil removal. J. Water Process Eng. **46**, 102643 (2022). https://doi.org/10.1016/j.jwpe.2022.102643
20. Ding, H., et al.: High hydrophobic poly(lactic acid) foams impregnating one-step Si–F modified lignin nanoparticles for oil/organic solvents absorption. Compos. Commun. **25**, 100730 (2021). https://doi.org/10.1016/j.coco.2021.100730
21. Xiao, W., et al.: Fabrication of foam-like oil sorbent from polylactic acid and *Calotropis gigantea* fiber for effective oil absorption. J. Clean. Prod. **278**, 123507 (2021). https://doi.org/10.1016/j.jclepro.2020.123507
22. Olivito, F., et al.: Bio-based polyurethane foams for the removal of petroleum-derived pollutants: sorption in batch and in continuous-flow. Polymers **15**(7), 1785 (2023). https://doi.org/10.3390/polym15071785
23. Lv, W., et al.: Superhydrophobic polyurethane foam based on castor oil and lignin with SiC nanoparticles for efficient and recyclable oil-water separation. J. Water Process Eng. **59**, 104897 (2024). https://doi.org/10.1016/j.jwpe.2024.104897
24. Vieira Amorim, F., José Ribeiro Padilha, R., Maria Vinhas, G., Ramos Luiz, M., Costa De Souza, N., Medeiros Bastos De Almeida, Y.: Development of hydrophobic polyurethane/castor oil biocomposites with agroindustrial residues for sorption of oils and organic solvents. J. Colloid Interface Sci. **581**, 442–454 (2021). https://doi.org/10.1016/j.jcis.2020.07.091

Chapter 5
Scalability and Other Considerations of Bio-based Foam Sorbents

Abstract Bio-based foam oil sorbents present a promising and sustainable approach for oil spill control. These sorbent materials have been widely studied for oil spill cleanup, remediation, and hazardous material containment. Each offers unique features regarding their compositions, syntheses, sorption performance, and practical manufacturing operations that render most impractical for larger oil spill applications. In addition, the nature of oil spills and the varying types of oils in terms of viscosity, density, polarity, and weathering may render these sorbents relatively more effective on a laboratory scale but underperform in actual and practical scenarios. This chapter outlines the challenges, strategies, and innovations necessary for advancing bio-based foam oil sorbents' large-scale adoption and commercialization, emphasizing the importance of balancing performance, cost, sustainability, and technological viability.

Keywords Oil spill control · Environmental impact · Bio-based sorbents · Sustainability · Scalability · Oil mobility · Surface chemistry · Commercialization · Structure optimization

5.1 Introduction

Oil spill control remains a significant environmental challenge, necessitating innovative and scalable solutions to mitigate the impacts of oil contamination-related accidents. Traditional techniques, such as skimming, in situ burning, and the application of chemical dispersants, have shown limited efficacy in fully containing and removing oil spills, particularly under complex marine conditions. The Deepwater Horizon oil spill in 2010 serves as a poignant example, revealing the shortcomings of these conventional methods. Despite extensive recovery efforts, substantial quantities of oil persisted in the marine environment, either dissolved or chemically altered, underscoring the urgent need for more effective and environmentally friendly technologies [1, 2].

A. A. Lubguban et al., *Bio-based Foam Sorbents*, Engineering Materials,
https://doi.org/10.1007/978-981-96-1460-8_5

Bio-based sorbents, derived from renewable resources, present a promising alternative for sustainable oil spill management. Unlike synthetic sorbents, which often involve complex manufacturing processes and non-biodegradable materials, bio-based options are more compatible with natural ecosystems and can potentially minimize ecological disruption [3, 4]. However, the scalability of these sorbents is a critical concern. Factors such as oil mobility, wave speed, oil viscosity, and environmental conditions must be thoroughly evaluated to determine the feasibility of deploying bio-based materials on a large scale [5]. Ongoing research aims to optimize the properties of these sorbents to enhance their performance, yet it is essential to assess their effectiveness against traditional materials under varying spill conditions [6, 7].

Traditional methods, including skimming and mechanical recovery, primarily focus on the physical containment and collection of oil. These methods are often employed alongside containment booms, which help concentrate oil spills for easier recovery. However, their effectiveness diminishes in rough sea conditions, limiting the efficiency of mechanical skimmers [8]. In contrast, bio-based sorbents may offer a more adaptable solution due to their lightweight nature and high surface area, which enhance oil absorption and retention [4, 9]. To fully understand their effectiveness, it is crucial to investigate how these materials interact with oils of varying viscosities and under different environmental stresses, as these factors directly influence recovery efficiency [10].

In situ burning is another widely utilized technique for oil spill mitigation, involving the ignition of oil on the water's surface to rapidly reduce its volume. While this method can be effective, it is not without drawbacks, including the release of toxic pollutants such as nitrogen oxides (NO_x) and sulfur oxides (SO_x), which pose risks to both marine and human health [11]. The success of in situ burning is contingent upon factors like oil thickness and weathering, which can significantly affect burn efficiency [9]. Bio-based sorbents may complement this method by acting as fire retardants or facilitating the removal of burn residues, indicating the need for further research into their multifunctional capabilities [4, 12].

Chemical dispersants, designed to break down oil slicks into smaller droplets, represent another conventional approach to oil spill control. While these chemicals enhance the natural degradation of oil by increasing its surface area, they do not eliminate the oil and can introduce long-term toxicity into marine environments, particularly in deep-sea ecosystems [13, 14]. In contrast, bio-based sorbents provide a more environmentally friendly solution by directly removing oil from water bodies without introducing additional pollutants [3, 4]. However, the efficiency of bio-based sorbents under real-world conditions—such as turbulent waters, varying temperatures, and mixed oil types—requires validation to ensure they can serve as a scalable alternative to chemical dispersants [15].

The scalability of bio-based sorbents hinges on their production and deployment at a reasonable cost while maintaining high performance. Current research has focused on optimizing the structure, surface chemistry, and porosity of these sorbents to maximize oil absorption and selectivity [4, 9]. Nevertheless, challenges such as durability, reusability, and the capacity to handle large-scale spills persist. This chapter outlines the challenges, strategies, and innovations necessary for advancing bio-based foam oil sorbents' large-scale adoption and commercialization, emphasizing the importance of balancing performance, cost, sustainability, and technological viability. Addressing these limitations will necessitate material innovation and a comprehensive understanding of how these sorbents perform in complex marine environments. Collaboration among academic researchers, industry experts, and environmental agencies is essential to develop bio-based sorbents that are both effective and economically viable [4, 6, 12].

5.2 Scalability

The scalability of bio-based oil sorbents is crucial for effective oil spill control. These eco-friendly alternatives to synthetic materials are derived from renewable sources and offer high absorption capacity. Scaling up involves refining raw materials, optimizing production, and ensuring cost-effectiveness to meet large-scale demands. Enhancing these sorbents regarding toxicity, properties, and reproducibility are key to their competitiveness.

5.2.1 Toxicity

Bio-based foam oil sorbents have gained attention due to their potential for effective oil spill cleanup while raising concerns about associated toxicities. These sorbents, often derived from renewable resources such as vegetable oils or natural fibers, can provide an environmentally friendly alternative to conventional petroleum-based sorbents. However, understanding their toxicological implications is crucial for ensuring safe application in environmental remediation. One significant concern regarding bio-based foam oil sorbents is potentially releasing toxic compounds during their production and degradation. For instance, Hoàng et al. highlighted that while their bio-polyurethane sorbent demonstrated high oil adsorption capacity and recyclability, the synthesis process involving chemical reactions could lead to the formation of by-products that may be harmful to aquatic life if released into the environment [16]. Similarly, using coconut oil-based polyols in polyurethane foams, as discussed by Tomon et al., raises questions about the thermal stability and potential leaching of toxic substances during the foam's lifecycle [17].

Moreover, the biodegradation of bio-based sorbents can also pose risks. While these materials are designed to be environmentally friendly, the breakdown products may not always be benign. For instance, the degradation of certain natural fibers used in sorbents can release microplastics or other harmful substances into the environment, although specific studies on this aspect are limited. Damanabi and Bahadori discussed challenges related to natural sorbents in oil spill cleanup, but their focus was primarily on the efficiency of oil absorption rather than degradation products

[18]. This highlights the need for thorough assessments of the degradation pathways and the environmental impact of these materials post-use.

Additionally, the incorporation of additives to enhance the performance of bio-based sorbents can introduce further toxicological concerns. For example, the study by Santos et al. on various polyurethane foams indicated that while additives like glycerol and polyethylene glycol can improve oil sorption capacity, they may also contribute to the overall toxicity of the sorbent material [19]. The balance between enhancing performance and maintaining environmental safety is a critical consideration in the development of these materials.

5.2.2 Properties

The commercialization of bio-based foam oil sorbents is significantly influenced by their properties, which encompass aspects such as oil adsorption capacity, biodegradability, mechanical strength, and overall environmental impact. These properties not only determine the effectiveness of the sorbents in oil spill cleanup but also influence market acceptance and regulatory compliance.

One of the critical factors affecting the commercialization of bio-based foam oil sorbents is their oil adsorption capacity. Research by Hoàng et al. indicates that the surface properties and porosity of bio-based polyurethane (PU) foams play a vital role in enhancing oil adsorption efficiency [16]. A highly porous material provides more interaction sites for oil, thereby increasing the overall oil removal efficiency [16]. This characteristic is essential for commercial viability, as products that demonstrate superior performance in oil absorption are more likely to gain traction in the market.

Mechanical properties, such as compressive strength, also play a crucial role in the commercialization of bio-based foam oil sorbents. Özbay and Yargıç noted that lower compressive strengths in bio-based foams could limit their applications, as they may not withstand the physical demands of oil spill cleanup operations [20]. Therefore, achieving a balance between cost and mechanical strength is essential for the successful commercialization of these materials. Products that maintain structural integrity while being environmentally friendly are more likely to succeed in the market.

5.2.3 Reproducibility

The reproducibility of bio-based sorbents for oil spill cleanup is a critical area of research, particularly in the context of increasing environmental concerns associated with oil spills. The reproducibility of these sorbent materials under varying conditions remains a challenge that must be addressed to ensure their practical application in real-world scenarios. Recent studies highlight the importance of material properties in determining the effectiveness of bio-based sorbents. For instance,

Herren et al. demonstrated that the incorporation of Joule heating functionality into sorbent materials can enhance oil absorption by reducing the viscosity of highly viscous oils, thereby improving the sorbent's performance in real-time applications [21]. This approach emphasizes the need for innovative designs that can adapt to the specific characteristics of oil spills, such as oil viscosity and environmental conditions. Furthermore, the structural properties of sorbents, including fiber diameter and porosity, significantly influence their oil retention capabilities [22]. These findings underscore the necessity for standardized testing protocols to evaluate the reproducibility of sorbent performance across different environmental conditions.

Reusability is another critical factor affecting the reproducibility of bio-based sorbents. Yadav et al. highlighted the challenges associated with the recycling of sorbent materials, particularly those derived from thermoplastic polystyrene, which often do not allow for effective recovery of absorbed oil [23]. The ability to reuse sorbents not only enhances their economic viability but also reduces the environmental footprint associated with their disposal. Therefore, developing bio-based sorbents that can be easily regenerated and reused is essential for ensuring long-term effectiveness in oil spill remediation. Moreover, the interaction between bio-based sorbents and various oil types must be thoroughly investigated to enhance reproducibility. Bhardwaj et al. reported that the performance of sorbents can vary significantly depending on the type of oil involved, with certain materials exhibiting superior absorption capabilities for specific oil viscosities [24]. This variability necessitates a comprehensive understanding of the physicochemical interactions between sorbents and oils, which can be achieved through systematic studies exploring oil properties' effects on sorbent performance.

In addition to material properties and oil interactions, environmental factors such as temperature, salinity, and wave action can influence the effectiveness of bio-based sorbents. For example, Chang et al. demonstrated that photothermal sponges could provide rapid cleanup of heavy oil spills under varying environmental conditions, suggesting that adaptability is crucial for the reproducibility of sorbent performance [25]. This adaptability can be further enhanced through the incorporation of advanced materials, such as carbon-based composites, which have shown promise in improving the reusability and, ultimately, the overall efficiency of oil absorption.

5.3 Sustainability

The increasing frequency and severity of oil spills worldwide have intensified the search for sustainable solutions to mitigate their environmental impacts. Among the various response strategies, bio-based sorbents have emerged as promising alternatives to traditional petroleum-derived sorbents due to their biodegradability and lower ecological footprint. This discussion explores the sustainability of bio-based sorbents, focusing on their biodegradability, global demand in suitability, cost-effectiveness, and environmental challenges associated with their end-of-life disposal in oil spill clean-up applications.

5.3.1 Biodegradability

The sustainability of bio-based sorbents for oil spill cleanup is increasingly recognized as a pivotal aspect of environmental remediation strategies. One of the key attributes that underpin the sustainability of these materials is their biodegradability. Unlike synthetic sorbents, which often persist in the environment and contribute to pollution, bio-based sorbents are derived from natural materials that can decompose and return to the ecosystem without leaving harmful residues. This characteristic is crucial in the context of oil spills, where the rapid and effective removal of contaminants is essential to minimize environmental damage.

Research has shown that various natural materials, such as agricultural waste and plant fibers, can be effectively utilized as bio-based sorbents. For instance, Zulkania et al. highlighted the potential of palm fiber waste as a sorbent for methylene blue, demonstrating that bio-sorbents derived from agricultural byproducts can effectively adsorb pollutants while being biodegradable [27]. This dual functionality not only aids in the cleanup of oil spills but also promotes waste valorization, contributing to a circular economy. The biodegradability of these materials ensures that they do not accumulate in marine environments, thereby reducing long-term ecological risks associated with oil spill remediation.

Moreover, the biodegradation process of bio-based sorbents is facilitated by various microbial communities in the environment. While Chakravarty's study focuses on the biodegradation of anthracene using Serratia quinivorans, it does not directly address the biodegradation of bio-based sorbents in oil spills [28]. However, it illustrates the potential of specific microorganisms to enhance the breakdown of organic pollutants, which can be relevant in oil spill remediation. The synergy between bio-sorbents and microbial degradation underscores the importance of integrating these materials with microbial processes to enhance the overall efficiency of oil spill cleanup operations.

The composition of bio-based sorbents plays a significant role in their biodegradability. Materials rich in cellulose, hemicellulose, and lignin, such as coir and kapok, have shown promising oil sorption capacities while being readily biodegradable [29]. The structural characteristics of these natural fibers facilitate microbial colonization and enzymatic breakdown, leading to efficient biodegradation. Additionally, the use of biopolymers like chitosan and alginate has been explored for their biodegradable properties, which can be tailored to enhance oil adsorption capabilities [30]. These advancements in material science highlight the potential for developing bio-based sorbents that are effective in oil spill cleanup and environmentally friendly.

However, the sustainability of bio-based sorbents is not without challenges. The variability in biodegradation rates among different materials can affect their performance in real-world scenarios. Factors such as environmental conditions, the presence of contaminants, and the specific microbial communities can influence the degradation process [31]. Therefore, it is essential to conduct comprehensive studies that evaluate the biodegradation kinetics of various bio-based sorbents under different environmental conditions to ensure their effectiveness in oil spill remediation.

5.3.2 Global Demand

The global demand for effective oil spill cleanup solutions has intensified in response to the increasing frequency and severity of oil spills, which pose significant environmental and economic threats. Traditional methods such as mechanical skimming and chemical dispersants have been widely employed, but they often fall short in efficiency and sustainability. In this context, bio-based sorbents have emerged as a promising alternative due to their biodegradable nature, cost-effectiveness, and potential for high oil absorption capacities. This discussion explores the sustainability of bio-based sorbents in addressing the global demand for oil spill remediation, highlighting their advantages, challenges, and future prospects.

Bio-based sorbents from renewable resources such as agricultural waste, natural fibers, and biodegradable polymers offer a sustainable approach to oil spill cleanup. Materials like kenaf, rice straw, and sawdust have been shown to possess excellent oil sorption capabilities, making them effective in absorbing oil from water surfaces [32]. Their biodegradability ensures that they do not contribute to long-term environmental pollution, a significant advantage over synthetic sorbents that may persist in ecosystems for extended periods [33]. Furthermore, using low-cost, abundant materials aligns with sustainability principles, reducing reliance on non-renewable resources and minimizing the carbon footprint associated with oil spill remediation efforts [34].

Despite their advantages, the scalability of bio-based sorbents poses challenges that must be addressed to meet global demand effectively. The production processes for many bio-based sorbents can be complex and costly, limiting their widespread adoption [35]. Additionally, the variability in the properties of natural materials can lead to inconsistent oil absorption performance, raising concerns about their reliability in real-world applications. Research has shown that optimizing the physical and chemical properties of bio-based sorbents, such as enhancing their hydrophobicity and oleophilicity, is crucial for improving their effectiveness and reproducibility [36]. Advances in material science, including the development of nanocellulose-based sorbents, have demonstrated potential for enhancing the performance of bio-based materials while maintaining sustainability.

The environmental impact of oil spills extends beyond immediate contamination; it also affects marine ecosystems and local economies reliant on fishing and tourism. Bio-based sorbents can play a vital role in mitigating these impacts by providing an effective means of oil recovery that minimizes ecological disruption. For instance, the use of biodegradable sorbents can facilitate the natural degradation of absorbed oil, promoting the restoration of affected habitats [37]. Moreover, the integration of bio-based sorbents with innovative technologies, such as photothermal fabrics that enhance evaporation rates, can further improve the efficiency of oil spill remediation efforts [38]. Such synergies highlight the potential for bio-based sorbents to contribute to a more sustainable and practical approach to oil spill cleanup.

As the global oil demand continues to rise, the risk of oil spills remains a pressing concern. The development of bio-based sorbents represents a proactive response to this challenge, offering a sustainable solution that aligns with environmental conservation goals. However, ongoing research and development are essential to overcome existing barriers, such as production scalability and performance consistency. Collaborative efforts among researchers, industry stakeholders, and policymakers will be crucial in advancing the adoption of bio-based sorbents and ensuring their integration into standard oil spill response protocols.

5.3.3 *Cost Effectiveness*

The sustainability of bio-based sorbents for oil spill cleanup is a critical area of research, particularly in light of the increasing frequency of oil spills and their environmental consequences. Bio-based sorbents from natural materials offer a cost-effective and environmentally friendly alternative to conventional synthetic sorbents. This discussion synthesizes recent findings on bio-based sorbents' effectiveness, cost-efficiency, and sustainability in oil spill remediation.

Bio-based sorbents are characterized by their biodegradability, low cost, and availability, making them attractive options for oil spill cleanup. Research has shown that materials such as rice husk, cotton, and kenaf fibers exhibit significant oil sorption capacities, which are essential for effective cleanup operations [33]. For instance, cotton products and kenaf have been highlighted for their hydrophobic and oleophilic properties, which enhance their ability to absorb oil [29]. Additionally, the use of agricultural waste products, such as rice husk and milkweed fibers, has been shown to provide effective oil absorption while being economically viable [33]. These materials not only reduce cleanup costs but also minimize the environmental impact associated with synthetic sorbents.

The cost-effectiveness of bio-based sorbents is further supported by their availability and the low processing costs associated with their production. Studies indicate that natural sorbents can be sourced from agricultural byproducts, which are often abundant and inexpensive [17, 26, 33]. For example, the use of waste chicken feathers and other biowastes has been explored as a sustainable approach to oil spill cleanup, demonstrating both economic and environmental benefits [39]. Moreover, the reusability of certain bio-based sorbents, such as those made from poly(ε-caprolactone) microfiber meshes, enhances their cost-effectiveness by allowing for multiple uses before disposal [40].

In terms of performance, bio-based sorbents have shown promising results in various studies. For instance, peat moss has been recognized for its high porosity and surface area, which contribute to its effectiveness as a hyper-sorbent for oil spills [41]. Additionally, the modification of natural fibers, such as through hydrophobic treatments, can significantly enhance their oil sorption capabilities [35]. These advancements in material science indicate that bio-based sorbents can be tailored to meet specific cleanup needs, further solidifying their role in sustainable oil spill remediation.

Despite the advantages, challenges remain in the widespread adoption of bio-based sorbents. Some studies have pointed out issues related to the mechanical strength and buoyancy of certain natural materials, which can limit their effectiveness in real-world applications [42]. Furthermore, while many bio-based sorbents are effective, the variability in performance based on material type and environmental conditions necessitates further research to optimize their use.

5.3.4 End-Of-Life Disposal

The end-of-life disposal of bio-based sorbents is undeniably noteworthy from environmental and economic points of view. Many of these materials are biodegradable, which means they can decompose naturally without causing long-term environmental harm. For example, cellulose and keratin-based biopolymeric sorbents have been highlighted for their eco-friendliness and potential to be integrated into a circular economy framework. This is crucial, as traditional synthetic sorbents often pose disposal challenges due to their non-biodegradable nature, leading to environmental pollution [43]. The biodegradability of bio-based sorbents allows for a more sustainable approach to oil spill clean-up, as they can be safely composted or converted into bioenergy after use, thereby minimizing waste.

However, the reusability of bio-based sorbents is another critical factor influencing their sustainability. Research indicates that some bio-sorbents can be treated and reused multiple times without significant loss of efficiency. For instance, modified waste cigarette filters have shown the ability to retain up to 75% of their sorption capacity after 20 cycles of reuse [34]. This characteristic not only enhances the economic viability of using bio-based sorbents but also reduces the frequency of disposal, further mitigating environmental impacts.

References

1. Zengel, S., et al.: Meta-analysis of salt marsh vegetation impacts and recovery: a synthesis following the *Deepwater Horizon* oil spill. Ecol. Appl. **32**(1), e02489 (2022). https://doi.org/10.1002/eap.2489
2. Mo, Y., Kearney, M., Riter, J.: Post-Deepwater horizon oil spill monitoring of Louisiana salt marshes using Landsat imagery. Remote Sens. **9**(6), 547 (2017). https://doi.org/10.3390/rs9060547
3. Da Silva, M.A., Soares, J.B.P.: Using poly(acrylamide-co-lauric acid) to remediate oil spills on water. Can. J. Chem. Eng. **101**(1), 322–327 (2023). https://doi.org/10.1002/cjce.24386
4. Yu, M., Zhu, Z., Chen, B., Cao, Y., Zhang, B.: Bioherder generated by *Rhodococcus erythropolis* as a marine oil spill treating agent. Front. Microbiol. **13**, 860458 (2022). https://doi.org/10.3389/fmicb.2022.860458
5. Klerks, P.L., Kascak, A., Cazan, A.M., Franco, M.E., Louka, F.R.: A mesocosm assessment of the effect of bioturbation by the Ghost Shrimp (*Lepidophthalmus louisianensis*) on the

fate of petroleum hydrocarbons in the intertidal zone. Environ. Toxicol. Chem. **39**(3), 637–647 (2020). https://doi.org/10.1002/etc.4652
6. Lyons, H.Z.: Responding to hard times in the 'big easy': meeting the vocational needs of low-income African American new Orleans residents. Career Dev. Q. **59**(4), 290–301 (2011). https://doi.org/10.1002/j.2161-0045.2011.tb00070.x
7. Kem, M.P., Zane, H.K., Springer, S.D., Gauglitz, J.M., Butler, A.: Amphiphilic siderophore production by oil-associating microbes. Metallomics **6**(6), 1150–1155 (2014). https://doi.org/10.1039/C4MT00047A
8. Ni, S., Qiu, W., Zhang, A., Prior, D.: Hydrodynamic simulation and optimization of an oil skimmer. J. Offshore Mech. Arct. Eng. **137**(2), 021301 (2015). https://doi.org/10.1115/1.4029211
9. Yang, S., Pu, F., Zhang, L., Li, M.: Boiling combustion behaviors and heat feedback of pool fire of diesel fuel–water emulsification. Process Saf. Prog. **43**(S1) (2024). https://doi.org/10.1002/prs.12554
10. Alloy, M., et al.: Ultraviolet radiation enhances the toxicity of *Deepwater Horizon* oil to Mahi-mahi (*Coryphaena hippurus*) embryos. Environ. Sci. Technol. **50**(4), 2011–2017 (2016). https://doi.org/10.1021/acs.est.5b05356
11. Goldman, Z.E., Kaufman, J.A., Sharpe, J.D., Wolkin, A.F., Gribble, M.O.: Coping with oil spills: oil exposure and anxiety among residents of Gulf Coast states after the Deepwater Horizon oil spill. UCL Open Environ. **4** (2022). https://doi.org/10.14324/111.444/ucloe.000035
12. Burggren, W., Dubansky, B., Roberts, A., Alloy, M.: Deepwater Horizon oil spill as a case study for interdisciplinary cooperation within developmental biology, environmental sciences and physiology. World J. Eng. Technol. **03**(04), 7–23 (2015). https://doi.org/10.4236/wjet.2015.34C002
13. Alloy, M., et al.: Co-exposure to sunlight enhances the toxicity of naturally weathered *Deepwater Horizon* oil to early lifestage red drum (*Sciaenops ocellatus*) and speckled seatrout (*Cynoscion nebulosus*). Environ. Toxicol. Chem. **36**(3), 780–785 (2017). https://doi.org/10.1002/etc.3640
14. Richardson, N.D., Macauley, M., Cohen, M.A., Anderson, R., Stern, A.: Managing risk through liability, regulation, and innovation: organizational design for spill containment in Deepwater drilling operations. Risk Hazards Crisis Publ. Policy **2**(2), 1–22 (2011). https://doi.org/10.2202/1944-4079.1083
15. Liu, Y., MacFadyen, A., Ji, Z.-G., Weisberg, R.H.: Introduction to monitoring and modeling the Deepwater Horizon oil spill. In: Liu, Y., MacFadyen, A., Ji, Z.-G., Weisberg, R.H. (eds.) Geophysical Monograph Series, vol. 195, pp. 1–7. American Geophysical Union, Washington, D. C. (2011). https://doi.org/10.1029/2011GM001147
16. Hoang, A.T., Nižetić, S., Duong, X.Q., Rowinski, L., Nguyen, X.P.: Advanced super-hydrophobic polymer-based porous absorbents for the treatment of oil-polluted water. Chemosphere **277**, 130274 (2021). https://doi.org/10.1016/j.chemosphere.2021.130274
17. Tomon, T.R.B., et al.: A novel naturally superoleophilic coconut oil-based foam with inherent hydrophobic properties for oil and grease sorption. Sci. Rep. **14**(1), 14223 (2024). https://doi.org/10.1038/s41598-024-64178-2
18. Damanabi, A.T., Bahadori, F.: Oil spill cleanup by atmospheric distillation. Pet. Sci. Technol. **34**(8), 719–725 (2016). https://doi.org/10.1080/10916466.2016.1163395
19. Santos, O.S.H., Da Silva, M.C., Yoshida, M.I.: Synthesis and performance of different polyurethane foams as oil sorbents. J. Appl. Polym. Sci. **134**(41), 45409 (2017). https://doi.org/10.1002/app.45409
20. Ozbay, N., Yargic, A.S.: Liquefaction of oak tree bark with different biomass/phenol mass ratios and utilizing bio-based polyols for carbon foam production. In: Presented at the Proceedings of the 6th International Advances in Applied Physics and Materials Science Congress & Exhibition: (APMAS 2016), p. 020039, İstanbul, Turkey (2017). https://doi.org/10.1063/1.4975454
21. Ghildiyal, V., Herel, R.V., Heffernan, B., Altaner, C.: The effect of Joule heating on collapse and water absorption of wood. Wood Mater. Sci. Eng. **18**(4), 1228–1236 (2023). https://doi.org/10.1080/17480272.2022.2121660

22. Kathir, K., Carroll, M.: From trash to treasure: a sustainable approach to oil spill clean-up. J. Emerg. Investig. (2023). https://doi.org/10.59720/21-018
23. Yadav, S., Mattaparthi, S., Sreenivasulu, K., Khandelwal, M., Majumdar, S., Sharma, C.S.: Recycling of thermoplastic polystyrene waste using citrus peel extract for oil spill remediation. J. Appl. Polym. Sci. **136**(33), 47886 (2019). https://doi.org/10.1002/app.47886
24. Bhardwaj, N., Bhaskarwar, A.N.: A review on sorbent devices for oil-spill control. Environ. Pollut. **243**, 1758–1771 (2018). https://doi.org/10.1016/j.envpol.2018.09.141
25. Chang, C.-Y., Lai, C.-L., Hwang, G.-J.: Trends and research issues of mobile learning studies in nursing education: a review of academic publications from 1971 to 2016. Comput. Educ. **116**, 28–48 (2018). https://doi.org/10.1016/j.compedu.2017.09.001
26. Zaro, M., Silvestre, W.P., Fedrigo, J.G., Zeni, M., Baldasso, C.: Sorption of oils by a commercial non-woven polypropylene sorbent. Res. Soc. Dev. **10**(14), e554101422671 (2021). https://doi.org/10.33448/rsd-v10i14.22671
27. Zulkania, A., Iqbal, M., Syamsumarlin, S.: Characterization of adsorbents derived from palm fiber waste and its potential on methylene blue adsorption. Key Eng. Mater. **841**, 273–277 (2020). https://doi.org/10.4028/www.scientific.net/KEM.841.273
28. Chakraborty, S., et al.: Self-activated energy release cascade from anthracene-based solid-state molecular solar thermal energy storage systems. Chem, p. S2451929424003127 (2024). https://doi.org/10.1016/j.chempr.2024.06.033
29. Anwana, E.E., Ewemoje, O.E.: Oil sorption performance of sorbent materials examined under static and dynamic conditions. Eur. J. Eng. Technol. Res. **6**(5), 107–110 (2021). https://doi.org/10.24018/ejeng.2021.6.5.2490
30. Chien, Y.-H., Lin, B.-Y., Shih, H.-H., Chen, C.-Y., Chen, P.-C.: Attracted-killed inhibition mechanism in Ag/chitosan hydrogel for long-acting control of *Ralstonia solanacearum*. Nanoscale, p. 10.1039.D4NR03306J (2024). https://doi.org/10.1039/D4NR03306J
31. Garg, S., et al.: Overview of natural source zone depletion: processes, controlling factors, and composition change. Groundw. Monit. Remediat. **37**(3), 62–81 (2017). https://doi.org/10.1111/gwmr.12219
32. Sultan Idris Education University, Malaysia, Shamsudin, R., Abdullah, H., the Department of Biology, Faculty of Science and Mathematics, Sultan Idris Education University, Malaysia, S. C. Sinang, and the Department of Biology, Faculty of Science and Mathematics, Sultan Idris Education University, Malaysia: Properties of oil sorbent material produced from Kenaf Fiber. Int. J. Environ. Sci. Dev. **6**(7), 551–554 (2015). https://doi.org/10.7763/IJESD.2015.V6.655
33. Doshi, B., Sillanpää, M., Kalliola, S.: A review of bio-based materials for oil spill treatment. Water Res. **135**, 262–277 (2018). https://doi.org/10.1016/j.watres.2018.02.034
34. Ifelebuegu, A.O., Johnson, A.: Nonconventional low-cost cellulose- and keratin-based biopolymeric sorbents for oil/water separation and spill cleanup: a review. Crit. Rev. Environ. Sci. Technol. **47**(11), 964–1001 (2017). https://doi.org/10.1080/10643389.2017.1318620
35. Oribayo, O., Feng, X., Rempel, G.L., Pan, Q.: Synthesis of lignin-based polyurethane/graphene oxide foam and its application as an absorbent for oil spill clean-ups and recovery. Chem. Eng. J. **323**, 191–202 (2017). https://doi.org/10.1016/j.cej.2017.04.054
36. Nandwana, V., Ribet, S.M., Reis, R.D., Kuang, Y., More, Y., Dravid, V.P.: OHM sponge: a versatile, efficient, and ecofriendly environmental remediation platform. Ind. Eng. Chem. Res. **59**(23), 10945–10954 (2020). https://doi.org/10.1021/acs.iecr.0c01493
37. Nazar, M., Ul Hassan Shah, M., Ahmad, A., Goto, M., Zaireen Nisa Yahya, W., Moniruzzaman, M.: Aggregation, toxicity, and biodegradability study of an ionic liquid-based formulation for effective oil spill remediation. Chemosphere **344**, 140412 (2023). https://doi.org/10.1016/j.chemosphere.2023.140412
38. Ku, Z., et al.: Molecular determinants and mechanism for antibody cocktail preventing SARS-CoV-2 escape. Nat. Commun. **12**(1), 469 (2021). https://doi.org/10.1038/s41467-020-20789-7
39. Osamor, A., Chinonyere, P.: Oil spill clean up from sea water using waste chicken feathers. In: Fourth International Conference on Advances in Applied Science and Environmental Technology—ASET 2016, Institute of Research Engineers and Doctors (2016), pp. 61–64. https://doi.org/10.15224/978-1-63248-097-2-42

40. Hersey, J.S., Yohe, S.T., Grinstaff, M.W.: Poly(ε-caprolactone) microfiber meshes for repeated oil retrieval. Environ. Sci. Water Res. Technol. **1**(6), 779–786 (2015). https://doi.org/10.1039/C5EW00107B
41. Pandey, S., Alam, A.: Peat moss: a hyper-sorbent for oil spill cleanup—a review. Plant Sci. Today **6**(4), 416–419 (2019). https://doi.org/10.14719/pst.2019.6.4.586
42. Viju, S., Thilagavathi, G.: Effect of Alkali treatment of nettle fibers on oil absorbency. J. Nat. Fibers **18**(12), 2092–2101 (2021). https://doi.org/10.1080/15440478.2020.1723776
43. Zamparas, M., Tzivras, D., Dracopoulos, V., Ioannides, T.: Application of sorbents for oil spill cleanup focusing on natural-based modified materials: a review. Molecules **25**(19), 4522 (2020). https://doi.org/10.3390/molecules25194522

Chapter 6
Future Directions and Innovations

Abstract The future of bio-based foam sorbents is marked by ongoing research and innovative developments aimed at enhancing their performance, expanding their applications, and facilitating their commercialization. This chapter reviews the latest research and development efforts focused on improving the properties of bio-based foam sorbents and integrating them with other sustainable materials. It also explores the potential for broader applications beyond oil spill treatment, highlighting innovations in related fields. Additionally, the chapter explores the commercialization of these materials, providing a market analysis, cost-benefit evaluation, and insights into industry involvement. The role of start-ups and research and development initiatives, along with the challenges and solutions associated with bringing bio-based foam sorbents to market, are also discussed.

Keywords Supercritical CO_2 · Surface functionalization · Grafting techniques · Market viability · Pollution control · Forestry residues · Composite materials · Hydrophobic agents · Non-isocyanate polyurethane (NIPU)

6.1 Research and Development of Bio-based Foam Sorbents for Oil Spill Containment

The research and development of bio-based foam sorbents have gained considerable momentum due to growing concerns over the environmental harm caused by traditional synthetic sorbents, which are often non-biodegradable and unsustainable. Bio-based sorbents, derived from renewable natural materials, present an eco-friendly alternative, effectively absorbing oil while minimizing environmental damage. A key area of focus in the development of these bio-based foams lies in enhancing their properties, particularly their hydrophobicity, oleophilicity, biodegradability, and reusability. These attributes are crucial for ensuring that bio-based sorbents are both efficient in oil spill containment and environmentally responsible [1].

A. A. Lubguban et al., *Bio-based Foam Sorbents*, Engineering Materials,
https://doi.org/10.1007/978-981-96-1460-8_6

6.1.1 Advancements in Bio-based Sorbent Properties

Recent advancements in bio-based foams have concentrated on optimizing foam structure to increase porosity and surface area, essential factors in maximizing oil retention. Supercritical CO_2 has emerged as an environmentally friendly foaming agent, creating interconnected porous structures that improve oil sorption capacity. This method also enhances the distribution of hydrophobic agents within the foam matrix, contributing to better oil absorption efficiency [2]. Furthermore, surface functionalization techniques, such as grafting hydrophobic groups onto foam surfaces, have been shown to increase selectivity and oil retention. Alkyl groups and fatty acid derivatives, for example, can improve the foam's resilience and performance in harsh maritime conditions [3, 4].

6.1.2 Optimizing Foam Structure for Enhanced Oil Absorption

The integration of nanofillers and natural fibers into bio-based foams has proven highly effective in enhancing sorbent properties. Nanofillers like zinc oxide (ZnO) nanoparticles significantly increase the foam's hydrophobicity, surface area, and adsorption capacity, while also improving mechanical strength and reusability. Coupling nanofillers with natural fibers, such as cellulose or agricultural waste fibers, results in composite materials that offer improved oil absorption and durability, providing a sustainable alternative to synthetic sorbents [5, 6]. These composite materials leverage the natural hydrophobicity and roughness of fibers, boosting oil sorption efficiency while contributing to a circular economy by utilizing waste materials.

6.1.3 Surface Functionalization for Improved Oil Retention

Surface functionalization plays a pivotal role in improving the selective oil absorption capabilities of bio-based foams. Grafting hydrophobic groups like alkyl chains or fatty acid derivatives onto the foam surface enhances the material's oleophilic nature, which is crucial for selective oil absorption. These modifications not only improve oil retention but also increase the durability of the sorbents under adverse environmental conditions such as oil spills in marine ecosystems [3, 4]. This research direction is key to making bio-based foams more resilient and reusable in harsh conditions.

6.1.4 Integration of Nanofillers and Natural Fibers in Bio-based Foams

The addition of natural fibers and nanofillers into bio-based foam matrices has been shown to significantly improve the performance of oil sorbents. Nanomaterials, such as zinc oxide (ZnO) nanoparticles, enhance the foam's hydrophobicity and mechanical strength, leading to increased oil adsorption capacity and reusability [6]. Moreover, coupling nanofillers with natural fibers like cellulose enhances the foam's surface roughness and structural integrity, which in turn increases oil absorption efficiency [5]. This integration represents a sustainable, efficient approach for improving the effectiveness of bio-based sorbents.

6.1.5 Utilization of Agricultural and Industrial Waste as Feedstock

Ongoing research also investigates the use of agricultural waste products as feedstock for bio-based sorbents. Materials like coconut shells, corn stalks, rice husks, and fruit peels have been identified as promising sources of cellulose and lignin, which can be used to produce foams with high oil absorption capacity. Utilizing these waste materials not only reduces reliance on synthetic foams but also addresses waste management challenges, aligning with global sustainability goals [7, 8]. Similarly, forestry residuals and industrial by-products like glycerol from biodiesel production are being explored as potential raw materials for bio-based foams [9].

6.1.6 Polyurethane and Non-isocyanate Polyurethane (NIPU) Foams

Polyurethane foams derived from vegetable oils, such as castor, soybean, and coconut oil, represent one of the most researched categories of bio-based foam sorbents. These foams exhibit excellent oil sorption capacities, hydrophobic characteristics, and biodegradability. Recent innovations in non-isocyanate polyurethane (NIPU) foams, which avoid the use of hazardous isocyanates, have garnered significant interest. NIPUs, synthesized from cyclic carbonates and amines, provide a safer and more environmentally friendly option for oil absorption applications [10, 11]. These developments not only enhance the environmental safety of foam production but also improve foam efficiency for use in oil spills and other applications.

6.1.7 Expanding Applications of Bio-based Sorbents Beyond Oil Spills

As the demand for sustainable environmental solutions grows, researchers are expanding the applications of bio-based foams beyond oil spill containment to address other types of environmental contaminants. By modifying foam structures and incorporating new feedstocks like palm oil, soybean oil, and algae, these sorbents can target a broader range of pollutants, including heavy metals and industrial chemicals. This expansion of functionality could make bio-based foams essential tools for broader environmental cleanup operations [12].

6.1.8 Challenges in Commercialization and Scalability of Bio-based Foams

As environmental regulations tighten and sustainability targets become more stringent, reducing production costs and improving scalability are crucial for making bio-based foams commercially viable alternatives to synthetic sorbents. This will require continued research into optimizing foam formulation, manufacturing processes, and the exploration of diverse bio-based feedstocks such as palm oil and algae, which could offer tailored solutions for different environmental challenges.

6.2 Broader Applications of Bio-based Foam Sorbents

Although bio-based foam sorbents were originally developed to tackle oil spills, their unique properties make them suitable for a wide range of industries and environmental applications. Their ability to selectively absorb hydrocarbons, oils, and greases makes them ideal for wastewater treatment, assisting industries in meeting stringent environmental regulations and preventing water contamination. In this application, bio-based foams effectively capture contaminants from industrial effluents, reducing pollutants discharged into water bodies and contributing to cleaner waterways [13]. This selective absorption capability is particularly advantageous in petrochemical and manufacturing industries where oil and grease-laden wastewater is a significant environmental challenge.

6.2.1 Applications in Agriculture for Soil Remediation

Beyond water treatment, bio-based foam sorbents can play a crucial role in agriculture, particularly in addressing soil contamination caused by the unintended release of petroleum-based chemicals or hazardous materials. Such contamination often results from accidental spills or the overuse of chemical products in agricultural practices. Bio-based foams offer a sustainable, long-term solution for soil remediation, efficiently absorbing harmful chemicals while maintaining biodegradability, which ensures that they do not introduce further pollutants into the ecosystem. Research has highlighted the potential for these foams to be used as part of a larger land rehabilitation strategy, ensuring soil health and reducing the environmental impact of contamination [14].

6.2.2 Potential in Food and Beverage Industry for Oil Filtration

Another promising area of application is in the food and beverage sector, where bio-based foams could be adapted for eco-friendly filtration processes. During food processing, oils and residues can accumulate, and traditional methods for removal often involve synthetic materials that pose environmental risks. Bio-based foams, with their ability to absorb oil selectively, provide a sustainable alternative for filtering out these unwanted residues. This innovation could help manufacturers reduce the environmental footprint of food production while maintaining high standards of cleanliness and product quality [15].

6.2.3 Multifunctional Use in Construction Industry

Bio-based foam sorbents also hold promise in the construction industry, where they could serve as multipurpose materials. Besides their potential as thermal insulators, these foams can absorb oil-based contaminants that may leak into infrastructure, preventing damage and contamination. By combining insulation properties with environmental protection, bio-based foams could enhance the sustainability of construction materials, making buildings both energy-efficient and environmentally friendly [16]. Additionally, the lightweight and porous nature of bio-based foams makes them well-suited for structural applications without compromising their absorption capabilities.

6.2.4 Eco-friendly Packaging Solutions

As global demand for sustainable packaging increases, bio-based foams are emerging as ideal candidates for green packaging solutions. Traditional packaging materials often contribute to waste and pollution, particularly when dealing with greasy or hazardous materials. Bio-based foams, however, provide a biodegradable option that not only cushions and protects products but also absorbs any oil or grease that might be released during shipping. This reduces the environmental impact of packaging waste, aligning with the global push for eco-friendly packaging alternatives [17].

6.2.5 Use in Emergency Response and Environmental Rehabilitation

In emergency response situations, such as oil spills or hazardous material leaks, bio-based foam sorbents offer rapid and effective cleanup without the negative environmental impacts associated with synthetic materials. These foams can be quickly deployed as absorbent pads or booms, containing and absorbing pollutants before they spread. As research advances, the development of bio-based foams capable of absorbing a wider range of industrial pollutants—such as heavy metals, solvents, and chemicals—could make them invaluable tools for comprehensive environmental rehabilitation projects [18]. This would position bio-based foams as essential assets in global efforts to address pollution in both land and water ecosystems.

The versatility of bio-based foam sorbents, from wastewater treatment to packaging and emergency response, underscores their potential to address some of the most pressing environmental challenges across multiple industries. As research continues to advance, the development of foams with specialized absorption capabilities could further expand their applications, making them integral to efforts in pollution control, environmental sustainability, and green technologies.

6.3 Commercialization of Bio-based Foam Sorbents

Commercializing bio-based foam sorbents presents both promising opportunities and notable challenges as the global market shifts toward sustainable environmental solutions. One of the primary obstacles to market penetration is the cost disparity between bio-based products and conventional petroleum-based sorbents. Despite their environmental benefits, bio-based foams must be financially viable to achieve widespread adoption. Achieving cost competitiveness requires scaling production and improving manufacturing efficiencies. This can be accomplished through innovations such as energy-efficient production techniques, the use of cheaper and more readily available raw materials, and leveraging economies of scale [19]. Collaboration between government agencies, private industry, and research institutions is essential to driving these innovations and reducing production costs. Governments can offer incentives, subsidies, or grants to stimulate the adoption of green technologies, while research institutions can focus on developing methods to enhance production efficiency.

6.3.1 Regulatory Drivers for Commercialization

The increasing regulatory pressure on industries to adopt sustainable practices—particularly in oil spill management and pollution control—creates a favorable environment for the commercialization of bio-based foam sorbents. Numerous countries are implementing stricter regulations regarding pollution control, oil spill response, and industrial waste management, prompting industries to seek eco-friendly alternatives. Bio-based foam sorbents present a strong commercial opportunity because they not only meet environmental sustainability goals but also comply with these emerging regulatory standards [20]. This regulatory shift, especially in sectors like oil and gas, is expected to drive demand for greener substitutes, with bio-based sorbents poised to benefit from the tightening legislation.

6.3.2 Reducing Production Costs and Enhancing Market Competitiveness

To make bio-based foams more commercially attractive, improving the cost-efficiency of production is paramount. Current research is focused on utilizing abundant, low-cost raw materials such as agricultural residues or industrial by-products, which can serve as feedstock for foam production. Additionally, optimizing production processes to reduce energy consumption and waste will be essential to reducing overall costs. Innovative techniques such as using supercritical CO_2 for foaming can not only enhance foam properties but also lower production costs by minimizing the need for synthetic chemicals or high-energy inputs [19]. The integration of renewable energy sources into the manufacturing process could further reduce costs and improve the sustainability profile of bio-based foam production, making it more competitive with petroleum-based alternatives.

6.3.3 End-of-Life Considerations and Circular Economy Integration

For bio-based foam sorbents to achieve commercial success, businesses must address their end-of-life disposal. Ensuring that these foams are either easily recyclable or fully biodegradable will increase their appeal, particularly to industries with a strong focus on environmental responsibility. By integrating these products into a circular economy model, manufacturers can emphasize their eco-friendly lifecycle, from production to disposal, reducing the overall environmental footprint [20]. Developing biodegradable or compostable foams that break down without leaving harmful residues is particularly important for sectors like packaging, construction, and environmental cleanup, where waste minimization is increasingly prioritized.

6.3.4 Market Potential and Consumer Appeal

A key aspect of successfully marketing bio-based foam sorbents is emphasizing their dual advantages: superior oil absorption performance and environmental sustainability. With the rising consumer demand for eco-friendly products, highlighting these benefits in marketing campaigns can attract a broad customer base, especially in sectors that prioritize green technologies. By showcasing the environmental impact reduction of bio-based foams, companies can tap into a growing market of environmentally conscious consumers [19]. Additionally, as sustainability becomes a core value for many businesses, bio-based foam sorbents may become the industry standard not just for oil spill response but also for broader applications in pollution control and industrial cleanup.

As industries increasingly focus on sustainability and regulatory compliance, bio-based foam sorbents are well-positioned to become a commercial success. By addressing cost challenges, improving production efficiency, and ensuring environmental responsibility throughout the product life-cycle, bio-based sorbents could emerge as a preferred choice in multiple sectors in the coming years.

References

1. Cui, S., et al.: Supercritical CO_2 extrusion foaming of highly open-cell poly(lactic acid) foam with superior oil adsorption performance. Int. J. Biol. Macromol. **269**, 132138 (2024). https://doi.org/10.1016/j.ijbiomac.2024.132138
2. Al Sumaiti, A., Shaik, A.R., Mathew, E.S., Al Ameri, W.: Tuning foam parameters for mobility control using CO_2 foam: field application to maximize oil recovery from a high-temperature high-salinity layered carbonate reservoir. Energy Fuels **31**(5), 4637–4654 (2017). https://doi.org/10.1021/acs.energyfuels.6b02595
3. Jarrah, K., Hisaindee, S., Al-Sayah, M.H.: Preparation of oil sorbents by solvent-free grafting of cellulose cotton fibers. Cellulose **25**(7), 4093–4106 (2018). https://doi.org/10.1007/s10570-018-1846-8
4. Yang, M., Gao, Y., Zhang, J., Zhou, F.: Hydrophobic modification of sepiolite by fatty acids for efficient oil removal: influence of the alkyl chain length. Appl. Clay Sci. **235**, 106867 (2023). https://doi.org/10.1016/j.clay.2023.106867
5. Hoàng, P.H., Dat, N.M., Hung, B.Q.: One-pot synthesis of bio-polyol from vegetable oil for preparation of bio-polyurethane based porous sorbent with high oil adsorption capacity. J. Appl. Polym. Sci. **140**(35) (2023). https://doi.org/10.1002/app.54332
6. Kovo, G., Akpomie, K., Conradie, J.: Enhanced surface properties, hydrophobicity, and sorption behavior of ZnO nanoparticle–impregnated biomass support for oil spill treatment. Environ. Sci. Pollut. Res. **28**(20), 25283–25299 (2021). https://doi.org/10.1007/s11356-021-12451-6
7. Valentukevičienė, M., Žurauskienė, R.: Investigating the effectiveness of recycled agricultural and cement manufacturing waste materials used in oil sorption. Materials **15**(1), 218 (2022). https://doi.org/10.3390/ma15010218
8. Elnaggar, E.: Synthesis of highly efficient sorbents from plant waste. Integr. Technol. Energy Saving **2**, 65–71 (2022). https://doi.org/10.20998/2078-5364.2022.2.06
9. Duc, N.T.: Sorption studies of crude oil on acetylated sawdust. Vietnam J. Sci. Technol. **54**(2A), 201 (2018). https://doi.org/10.15625/2525-2518/54/2a/11931

10. Tomon, R.B.T., et al.: A novel naturally superoleophilic coconut oil-based foam with inherent hydrophobic properties for oil and grease sorption. Sci. Rep. **14**(1), 14223 (2024). https://doi.org/10.1038/s41598-024-64178-2
11. Rayung, M., Ghani, N.A., Hasanudin, N.: A review on vegetable oil-based non-isocyanate polyurethane: towards a greener and sustainable production route. RSC Adv. **14**, 9273–9299 (2024). https://doi.org/10.1039/d3ra08684d
12. Tan, J.Y., Low, S.Y., Ban, Z.H., Siwayanan, P.: A review on oil spill clean-up using bio-sorbent materials with special emphasis on utilization of kenaf core fibers. BioResources **16**(4), 8394–8416 (2021). https://doi.org/10.15376/biores.16.4.8394-8416
13. Wu, S., Shi, W., Li, K., Cai, J., Chen, L.: Recent advances on sustainable bio-based materials for water treatment: fabrication, modification and application. J. Environ. Chem. Eng. **10**(6), 108921 (2022). https://doi.org/10.1016/j.jece.2022.108921
14. Osman, A.I., Farghali, M., Rashwan, A.K.: Life cycle assessment of biochar as a green sorbent for soil remediation. Curr. Opin. Green Sustain. Chem. **46**, 100882 (2024). https://doi.org/10.1016/j.cogsc.2024.100882
15. D'Souza, G., et al.: Functional bio-foam filters: an effective barrier for microplastic and other emerging pollutant containment. Mater. Today Sustain. (2024). https://doi.org/10.1016/j.mtsust.2024.100777
16. Olivito, F., et al.: Bio-based polyurethane foams for the removal of petroleum-derived pollutants: sorption in batch and in continuous-flow. Polymers **15**(7), 1785 (2023). https://doi.org/10.3390/polym15071785
17. Wang, X., et al.: Starting materials, processes and characteristics of bio-based foams: a review. J. Bioresour. Bioprod. **9**(2), 160–173 (2024). https://doi.org/10.1016/j.jobab.2024.01.004
18. Pattanayak, R., Jena, T.: Bio-based superabsorbent polymer: Current trends, applications and future scope. Eng. Mater. 185–197 (2023). https://doi.org/10.1007/978-981-99-3094-4_10
19. Gote, H.H., Martin, D., Muley, S.: Advanced synthetic and bio-based sorbents for oil spill clean-up: a review of novel trends. Nat. Environ. Pollut. Technol. **22**(1), 39–61 (2023). https://doi.org/10.46488/nept.2023.v22i01.004
20. Kasbaji, M., et al.: Bio-based functionalized adsorptive polymers for sustainable water decontamination: a systematic review of challenges and real-world implementation. Environ. Pollut. 122349 (2023). https://doi.org/10.1016/j.envpol.2023.122349

MIX
Papier aus verantwortungsvollen Quellen
Paper from responsible sources
FSC® C105338

If you have any concerns about our products,
you can contact us on
ProductSafety@springernature.com

In case Publisher is established outside the EU,
the EU authorized representative is:
Springer Nature Customer Service Center GmbH
Europaplatz 3, 69115 Heidelberg, Germany

Printed by Libri Plureos GmbH
in Hamburg, Germany